AF615623

THE SQUIBB INSTITUTE FOR MEDICAL RE
LIBRARY
OCT 1 5
PRINCETON, N. J.

Steroid Biochemistry

Volume I

Editor

Ronald Hobkirk, Ph.D., D. Sc.

Professor of Biochemistry
University of Western Ontario
London, Ontario, Canada

CRC PRESS, INC.
Boca Raton, Florida 33431

Library of Congress Cataloging in Publication Data

Main entry under title:

Steroid biochemistry.

Includes index.
1. Steroid hormone metabolism. 2. Biological chemistry. I. Hobkirk, Ronald, 1930- [DNLM: 1. Steroids — Biosynthesis. 2. Steroids — Metabolism. QU85 S8355]
QP572.S7S73 599'.01'9243 79-11988
ISBN 0-8493-5193-6 (v. 1)
ISBN 0-8493-5194-4 (v. 2)

This book represents information obtained from authentic and highly regarded sources. Reprinted material is quoted with permission, and sources are indicated. A wide variety of references are listed. Every reasonable effort has been made to give reliable data and information, but the author and thc publisher cannot assume responsibility for the validity of all materials or for the consequences of their use.

All rights reserved. This book, or any parts thereof, may not be reproduced in any form without written consent from the publisher.

© CRC Press, Inc., 1979

International Standard Book Number 0-8493-5193-6 (Volume I)
International Standard Book Number 0-8493-5194-4 (Volume II)
Library of Congress Card Number 79-11988
Printed in the United States

PREFACE

It was initially anticipated that about eight or nine chapters on various topics in steroid biochemistry would constitute a single volume of this publication. However, it became apparent that the literature in certain of the areas covered was so considerable as to necessitate two volumes, each containing four chapters. Since it was clearly advantageous that these volumes should be of approximately equal length, the original order of chapters had to be altered somewhat. This change, however, has not resulted in any major disadvantage.

An attempt has been made to attract contributions which illustrate the importance of certain enzymatic processes involved in steroid biosynthesis and metabolism and, in some cases, leading to steroidal action in target sites. Investigators actively engaged in research in such areas were invited to present their material in a manner which they considered fitting. It is hoped that as a result of this, the publication will possess sufficient depth to warrant approval. The blend of review material and experimental data originating in the authors' laboratories will, it is felt, make for useful reading.

As is the case with virtually every publication which involves multiple contributors, the various chapters were completed at rather different times, resulting in some being more current than others. For this reason dates of receipt are included in the following brief description of the contents.

In Volume I the chapter by A. F. Clark (July 1978) deals with the reductase enzymes acting at carbon-5 of the steroid molecule, with particular emphasis upon the potential hormonal activities of the products. An in-depth view of the chemistry and biochemistry of steroid carboxylic acids is provided by H. L. Bradlow and C. Monder (November 1977). It should be noted that the original work of these authors provided most of our current knowledge on this topic. D. G. Williamson's review (June 1977) of the biochemistry of 17-hydroxysteroid dehydrogenases focuses upon the role of these enzymes in the biological activation and inactivation of steroids. A. H. Payne and S. S. Singer (July 1977) discuss the role of steroid sulfotransferase and sulfatase systems from a biochemical viewpoint. In addition, they draw attention to the possible biological role of glucocorticoid sulfates. Volume II of the publication contains a chapter by B. R. Bhavnani and C. A. Woolever (August 1977) on biosynthetic pathways and some aspects of metabolism and biological activity of ring B unsaturated estrogens. This is a topic which has been primarily elucidated by the experimental work of Bhavnani. J-G. Lehoux (July 1977) reviews the complex area of the control of mineralocorticoid biosynthesis and the involvement of peptide hormones, electrolytes, prostaglandins, serotonin, etc. P. I. Musey, K. Wright, J. R. K. Preedy and D. C. Collins (July 1978) deal with steroid conjugate formation in relation to steroid distribution in body tissues and fluids. Factors related to the biological hydroxylation of 18- and 19-carbon steroids are discussed in a chapter by R. Hobkirk (May 1977).

It is my distinct pleasure to thank each contributor for the effort put into this publication. I should also like to acknowledge the help accorded me by CRC Press, particularly by Coordinating Editor Benita Segraves.

R. Hobkirk
London, Canada

THE EDITOR

Dr. R. Hobkirk was born in the town of Peebles in Scotland and attended the local High School. After graduating B.Sc. with Honors in Biochemistry from the University of Edinburgh in 1952 he completed a Ph.D. degree at the same institution in 1955. This latter involved studies on the biochemistry of plant cell wall polysaccharides. The following two years were spent as a research fellow (British Empire Cancer Campaign) in the Departments of Biochemistry and Surgery, University of Glasgow. During that period studies were performed on the relationship of steroid hormones to human breast cancer. Part of the 1957—'58 year was spent as a research fellow in the Department of Biochemistry, McGill University, Montreal and then, until 1960, in the Department of Metabolism, The Montreal General Hospital, Montreal. In 1960 Dr. Hobkirk was appointed Assistant Professor in Experimental Medicine, McGill University, followed in 1962 by promotion to Associate Professor and in 1967 to Professor (Biochemistry) in the Department of Medicine. In 1966 he was appointed a Research Associate of the Medical Research Council of Canada (i.e., a full time investigatorship), an award which he continues to hold. In 1971 he moved to the University of Western Ontario, London, Canada, and is currently Professor in the Department of Biochemistry. He is occupied with research and teaching in hormone biochemistry. In 1973 he received the degree of D.Sc. from the University of Edinburgh for a thesis entitled, "Metabolism of the Estrogens and their Conjugates".

CONTRIBUTORS

Bhagu R. Bhavnani, Ph.D.
Associate Professor
Department of Obstetrics and Gynecology
McMaster University
Hamilton, Ontario, Canada

H. Leon Bradlow, Ph.D.
The Rockefeller University
New York City, New York

Albert F. Clark, Ph.D.
Professor of Biochemistry and Associate Professor of Pathology
Queen's University
Kingston, Ontario, Canada

Delwood C. Collins, Ph.D.
Professor of Medicine
Associate Professor of Biochemistry
Emory University School of Medicine
Atlanta, Georgia

Ronald Hobkirk, Ph.D., D. Sc.
Professor of Biochemistry
University of Western Ontario
London, Ontario, Canada

Jean-Guy Lehoux, Ph.D.
Associate Professor, Faculty of Medicine
Sherbrooke University
Sherbrooke, Quebec, Canada

Carl Monder, Ph.D.
Professor of Biochemistry
Mt. Sinai School of Medicine
Director, Section on Steroid Studies
Hospital for Joint Diseases
New York, New York

Paul I. Musey, Ph.D.
Assistant Professor of Medicine
Emory University School of Medicine
Atlanta, Georgia

Anita H. Payne, Ph.D.
Associate Professor of Biological Chemistry
Department of Obstetrics and Gynecology
University of Michigan
Ann Arbor, Michigan

John R. K. Preedy, M.D.
Professor of Medicine,
Associate Professor of Biochemistry
Emory University School of Medicine
Atlanta, Georgia

Sanford S. Singer, Ph.D.
Associate Professor of Chemistry
University of Dayton
Dayton, Ohio

Denis G. Williamson, Ph.D.
Associate Professor
Department of Biochemistry
University of Ottawa
Ottawa, Ontario, Canada

Charles A. Woolever, M.D.
Professor of Obstetrics and Gynecology
McMaster University
Hamilton, Ontario, Canada

Kristina Wright, Ph.D.
Instructor of Medicine
Emory University School of Medicine
Atlanta, Georgia

TABLE OF CONTENTS

Volume I

Volume II

Chapter 1

STEROID Δ^4-REDUCTASES: THEIR PHYSIOLOGICAL ROLE AND SIGNIFICANCE

Albert F. Clark

TABLE OF CONTENTS

I. INTRODUCTION

In this review of the steroid Δ^4-reductases, the enzymes which catalyze the reduction of the double bond between carbons 4 and 5 of the steroid molecule, only mammalian steroids and mammalian enzymes will be discussed. The literature in this area of steroid metabolism is very voluminous and it will not be possible to quote all references on the topic. The general aim of the review is to indicate the physiological importance of Δ^4-reduction in the overall disposition of steroid hormones and, in appropriate situations, how it affects biological actions of the parent compounds. A few general comments on the biochemistry of the enzymes will be made in this introductory section.

With the exception of the estrogens, which have an aromatic ring A, the other biologically important, naturally occurring mammalian steroid hormones have in general the Δ^4-3-ketone structure in ring A. This group includes the important glucocorticoids cortisol and corticosterone, the mineralocorticoid aldosterone, the androgen testosterone, and the progestin progesterone. Key biosynthetic precursors of these hormones (such as 11-deoxycortisol, deoxycorticosterone, 18-hydroxycorticosterone, and Δ^4-androstenedione) also have the Δ^4-3-ketone configuration in ring A. Δ^4-3-Keto-intermediates are involved in the conversion of cholesterol to bile acids.

The reduction of the double bond between carbons 4 and 5 of steroid molecules can either be the first step in a sequence of reactions leading to water-soluble conjugates which are excreted or result in the formation of compounds of key biological importance. The reduction of the double bond between steroid carbons 4 and 5 involves the addition of two hydrogen atoms. Because the reduction leads to the formation of an assymetric carbon atom at carbon 5, either of two possible isomers can result. By

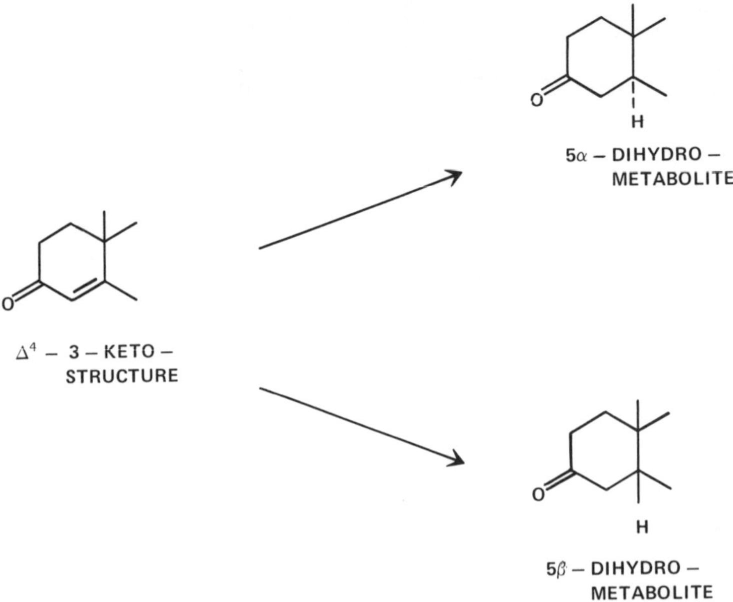

FIGURE 1. Steroid ring A reduction pathways yielding 5α- and 5β-metabolites.

standard steroid nomenclature these are the 5α- (hydrogen below plane of steroid molecule) and the 5β- (hydrogen above plane of steroid molecule) reduction products (see Figure 1). Both isomers are formed in mammals.

As mentioned above, the enzymes catalyzing these reductions are called Δ^4-reductases and can be either Δ^4-5α-reductases or Δ^4-5β-reductases, depending on the stereochemistry of the product at carbon 5. The Δ^4-reductases catalyze an irreversible reduction as they are not capable of removing hydrogens from positions 4 and 5 of saturated steroids.

There have been two common approaches to the measurement of Δ^4-reductase activity. Steroids with a conjugated double bond system in ring A (Δ^4-3-ketone) have a characteristic absorption peak in the ultraviolet range in organic solvents at or close to 240 nm. This absorption peak disappears when the double bond between carbons 4 and 5 is reduced and, hence, this has provided the basis for many Δ^4-reductase assays, particularly in the liver. The assay does not indicate in any way whether the product formed is a 5α- or a 5β-reduced steroid. However, in tissues other than liver, the rate of reduction is often not great enough to follow with changes in ultraviolet absorption spectra and ^{3}H- or ^{14}C-labeled steroids are utilized as substrates; the products are then isolated by standard chromatographic techniques and the rate of formation of the labeled products is used as the basis of assays for measuring Δ^4-reductase activity.

The required cofactor for all these Δ^4-reductases is NADPH. Bjorkhem[1] examined the mechanism and stereochemistry of rat liver microsomal Δ^4-5α-reductase, utilizing 7α-hydroxycholest-4-en-3-one as substrate, and found the hydrogen to be preferentially transferred from the 4B position of NADPH. The evidence obtained suggested a *cis* or nonspecific addition of hydrogen rather than a *trans* addition. Bjorkhem and Danielsson[2] then investigated rat liver soluble Δ^4-5β-reductase and found, using C_{27}-, C_{24}-, C_{21}-, and C_{19}-steroid substrates, that the 4A hydrogen of NADPH was transferred during the reduction reaction. Abul-Hajj[3] confirmed the specificity of the 4B and 4A hydrogen transfer for the rat liver Δ^4-5α-reductase and Δ^4-5β-reductases, respectively, using testosterone as substrate.

When the details of Δ^4-reductases for each of the hormone groups are examined, one notes that in general the liver contains both Δ^4-5β-reductase and Δ^4-5α-reductase activities while other tissues have only Δ^4-5α-reductase activity. There are a few exceptions but in general there is no known biological significance to the existence of Δ^4-5β-reductases is tissues other than the liver. For most steroids the liver is the principal site for their metabolism. There is considerable evidence that Δ^4-reduction is the rate-limiting step in steroid metabolism and this is supported by the finding of low levels of 5α- and 5β-reduced 3-ketosteroids in the circulation. In other words, once the Δ^4-reductases have performed their actions, other enzymes, in particular the 3-hydroxysteroid dehydrogenases, reduce the ring A saturated steroids to 3α- and 3β-hydroxysteroids, which in turn are then substrates for β-glucuronyl transferases and sulfokinases.

One other generalization for which examples will be noted later is that in the liver, Δ^4-5β-reductases are soluble enzymes, i.e., they are found in the 105,000 × g cell homogenate supernatant, while the Δ^4-5α-reductases are located in the microsomes or endoplasmic reticulum. In other tissues the Δ^4-5α-reductases are also located in the microsomes and in the nuclei in some instances as well.

Details on the steroid Δ^4-reductases will be discussed for the glucocorticoids (cortisol and corticosterone), aldosterone, progesterone, and C_{19}-steroids. Those involved in bile acid metabolism will not be discussed. There have been studies on the formation of Δ^4-reduced steroid metabolites in many species. Many in vivo studies have been conducted in humans, but these can be in some ways considered to be indirect in that the urine metabolites are used as indicators of the extent and type of Δ^4-reduction. These studies are complemented in many instances by in vitro experiments with tissues, tissue fractions, and purified enzymes from animals. Such a situation is of course by necessity since the appropriate human tissues are often not readily available, especially in experimental situations where changes in enzyme levels are being followed. However, one must also remember the possibility of species differences when results for humans and animals are being compared. Because of the many studies in the literature on the use of urine steroid metabolite levels to indicate or follow changes in steroid secretion by the adrenal cortex or gonads, examples of such will be given in greater detail for the corticoids since Δ^4-reduction is not believed to be greatly involved in their actions. Similar principles apply to the androgens and progestins with regard to steroid metabolites and hence fewer details will be given for them. However, the involvement of Δ^4-5α-reduction in the action of progestins and androgens will be detailed more for the latter two groups of steroids because its importance has been better documented; this applies in particular to the androgens.

II. CORTICOIDS

This section will deal with some of the products secreted by the adrenal cortex, in particular, glucocorticoids and mineralocorticoids. Not all related steroids will be mentioned as in some instances they are precursors of the biologically important compounds and they are all metabolized similarly in ring A. Cortisol, being the principal glucocorticoid in humans, has been the subject of a number of studies with regard to its metabolism. Until very recently no physiological significance has been attached to the Δ^4-reduction of cortisol other than its role in the pathway leading to metabolites completely reduced in ring A which can be conjugated with glucosiduronic acid and excreted in the urine. Hence, the Δ^4-reducion of cortisol is not involved directly in the expression of corticoid activity. The effects will only be indirect in that altered rates of liver Δ^4-reduction can lead to changes in the amount of hormone available for the expression of activity. Recently, a case of hypertension with evidence of increased mineralocorticoid production has been reported in which the abnormality found in steroid metabolism was an increased excretion of unconjugated dihydrocortisol. This has stimulated some studies on the possible biological significance of such metabolites. There is also some indirect evidence for decreased Δ^4-5β-reduction of aldosterone in certain cases of hypertension, resulting in a decreased rate of clearance and a resulting increased expression of activity of this potent mineralocorticoid.

Cortisol is metabolized in large part by the liver although other tissues have a capacity, albeit small, to perform some of the metabolic conversions. Very little of the total cortisol produced is excreted unchanged (approximately 100 μg from a total of 20 mg); this urine excretion has been used as an indicator of the free or unbound fraction of serum cortisol.[4] Perfusion of a human liver with labeled cortisol led to significant amounts of ring A reduced metabolites in the perfusate during the first 5 min.[5] Metabolites in the unconjugated fraction tended not to be reduced in ring A supporting the requirement for these reductions prior to conjugation. In the liver cortisol is metabolized via both Δ^4-5α-reduction and Δ^4-5β-reduction with the resulting dihydrocortisols being further reduced to tetrahydrocortisols (see Figure 2).[6] The major tetrahydrocortisol excreted in the urine has the 3α-hydroxy-5β-stereoconfiguration; the corresponding 5α-isomer is excreted to a much lesser extent. These products make up a significant percentage of total neutral metabolites of cortisol.[7] Most of the other cortisol metabolites excreted are also reduced in ring A. They include tetrahydrocortisones (11β-hydroxyl of cortisol oxidized to 11-ketone), cortols (ketone at carbon-20 reduced to hydroxyl), carboxylic acid metabolites (side chain of cortisol oxidized), and C_{19}-steroids

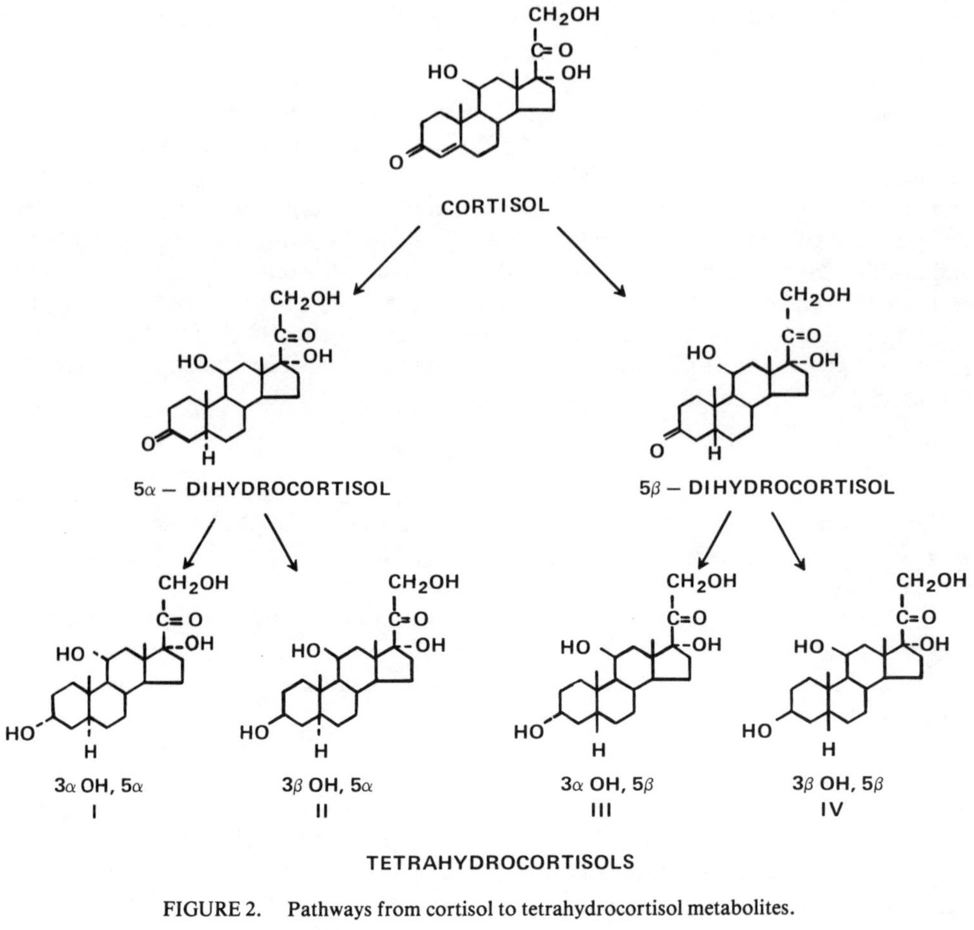

FIGURE 2. Pathways from cortisol to tetrahydrocortisol metabolites.

(cortisol side chain cleaved). These metabolites of cortisol for humans have been the subject of a number of publications.[7,8] Of all these various cortisol metabolites, most are 5β-products. Ring A reduction does not mean the end of cortisol metabolism; injection of 4-^{14}-C-tetrahydrocortisol to normal men led to the excretion of significant amounts of a number of labeled metabolites including tetrahydrocortisone, cortol, and 11β-hydroxyetiocholanolone.[9] Significantly, no labeled 5α-reduced metabolites could be detected, supporting the concept of the irreversibility of Δ^4-reduction.

The urine 3α-hydroxy-5β-metabolites of both cortisol and cortisone, i.e., tetrahydrocortisol and tetrahydrocortisone, have been considered unique metabolites of cortisol; that is, they are derived solely from cortisol. Using this assumption, the specific activity of one of these urine metabolites following an injection of ^{3}H- or ^{14}C-labeled hormone has been utilized to calculate the cortisol production rate.[10] Results must at times be interpreted with caution because of some possible irreversible metabolism leading to tetrahydro metabolites occurring prior to mixing of injected with endogenous cortisol.[11] As critically reviewed,[12] production rates based on the use of specific activities of urine tetrahydrocortisol and tetrahydrocortisone following the intravenous administration of labeled cortisol may differ; the reasons for this are not clear, but the irreversible metabolism prior to mixing and liver compartmentation may be involved.

There is evidence in the rat liver for a series of microsomal Δ^4-5α- and soluble Δ^4-5β-reductases differing in substrate specificities with one at each subcellular site being relatively specific for cortisol;[13,14] another may exist for the closely related cortisone. Studies on factors which influence cortisol Δ^4-reduction have, in general, been performed on animal livers although there are reports of indirect studies whereby human urine and plasma metabolites are examined. In the rat, corticosterone rather than cortisol is, of course, the predominant glucocorticoid. Estrogen increases the activity of the liver corticosterone Δ^4-reductase activity in male rats but decreases it in female rats.[15] Testosterone administration decreases liver corticosterone Δ^4-reductase in orchidectomized male rats.[16] The presence of the pituitary is required for the above effects of estradiol and testosterone, indicating that overall control of liver Δ^4-reductase activity is complex.[16] However, estrogen also causes increases in plasma transcortin levels, which has a significant physiological impact on glucocorticoid clearance. Transcortin binds cortisol, decreasing its clearance from the circulation by preventing its entry into tissues where it can be metabolized.[17] Theoretically, the increased liver Δ^4-5α-reductase level would increase liver clearance of cortisol, but this is difficult to measure independently of changes in the transcortin level. Heparin, when administered in large quantities, decreases the liver corticoid Δ^4-reductase activity.[18] In vitro evidence suggests that heparin may compete with the cofactor NADPH for the enzyme. It was found that heparin injections significantly prolonged the half-life of ^{14}C-corticosterone. Also, heparin had an anti-inflammatory effect in the intact animal but not in the adrenalectomized one unless cortisol was also administered. This supports the contention that alterations in liver Δ^4-reductase activity will alter corticoid metabolism in vivo and hence influence the biological activity of the steroids.

ACTH administration to humans decreases the formation of Δ^4-reduced metabolites of injected cortisol-4-^{14}C;[19] this at least in part accounts for the resulting prolonged cortisol half-life. The site of this effect can only be presumed to be the liver since it is the major site of cortisol metabolism. While thyroxine administration to rats increases Δ^4-5α-reductase activity for both cortisol and testosterone,[20] the altered states of thyroid function in humans affects only testosterone metabolism, there being no alterations in the ratio of urine 5α- to 5β-metabolites of cortisol.[21]

Until the above-mentioned report[22,23] that a hypertensive patient excreted increased amounts of free dihydrocortisol with the 5α-isomer predominating, relatively little physiological activity was associated with the reduced metabolites of cortisol. This report led to an investigation of dihydrocortisol as a potential mineralocorticoid.[24] It was found in rat kidney slices that 5α-dihydrocortisol and 5β-dihydrocortisol competed with aldosterone for its receptor but not with dexamethasone for the glucocorticoid receptor. Because neither bind strongly with transcortin they would be free to enter target tissues if present in the circulation in significant amounts. In concentrations 500 times greater than aldosterone, 5α-dihydro- but not 5β-dihydrocortisol significantly stimulated sodium transport in the isolated toad bladder. A similar effect was found when the urine K/Na ratio was followed in adrenalectomized rats infused with the steroids, indicating that 5α-dihydrocortisol had effects similar to those of aldosterone in this mineralocorticoid assay system as well, i.e., it stimulated sodium reabsorption and potassium excretion. Hence, it was concluded that 5α-dihydrocortisol could be a significant mineralocorticoid in certain situations.

Several situations exist where the extent of cortisol being metabolized to tetrahydro metabolites is decreased with a corresponding increase in production of other metabolites such as 6β-hydroxycortisol, which is more polar and can be excreted in a free or unconjugated state. Examples of such situations include pregnancy,[25] neonatal life,[26] Cushing's syndrome,[27] and during the administration of drugs such as diphenylhydantoin and "o,p′-DDD" (2,2-bis (2-chlorophenyl-4-chlorophenyl)-1,1-dichloroethane).[28,29]

Several other tissues have the capacity to reduce ring A of glucocorticoids although quantitatively they may not make an important overall contribution in vivo. These studies in some instances have involved glucocorticoid target tissues. The rat lung has been reported[30] to have Δ^4-5α-reductase activity for cortisone. Chick embryo bone in culture[31] metabolized cortisol to 5β-dihydrocortisol and tetrahydrocortisol; only the substrate inhibited bone formation with the metabolites having no effect. No evidence of cortisol ring A reduction was found when it was injected in vivo into the knee synovial cavity although other metabolites were found in the synovial fluid 15 min later.[32] The kidney does not appear to have significant glucocorticoid Δ^4-reductase activity.[33]

There have been a number of interesting studies from the laboratories of Colby and Kitay on corticosterone Δ^4-5α-reductase in the rat adrenal cortex. This enzyme is under the control of a number of hormones. There is a relationship between the levels of this enzyme and the secretory products of the adrenal cortex, e.g., when enzyme activity is high corticosterone secretion decreases and the rates of 5α-dihydrocorticosterone and 3β,5α-tetrahydrocorticosterone secretion increase. Hypophysectomy leads to an increased adrenal 5α-reductase activity in 1 week and this response appears to play a significant role in the decreased responsiveness of the adrenal to ACTH when corticosterone secretion is the measured end point.[34] Several days of ACTH administration leads to decreased enzyme activity and increased corticosterone secretion in response to ACTH injection.[34] Administration of growth hormone and prolactin as well as the steroid hormone testosterone and estradiol all lead to decreases in corticosterone Δ^4-5α-reductase activity.[35,36] The effects of several of these hormones are probably additive. The pineal gland is also involved in the control of the adrenal Δ^4-5α-reductase; removal of the pineal leads to a decrease in enzyme activity and a decrease in the secretion of reduced corticosterone metabolites.[37] The effect appears to be a direct one of melatonin rather than one that is mediated by the adenohypophysis as direct addition of this compound to adrenal slices stimulates Δ^4-5α-reductase activity.[38] The levels of corticosterone in the circulation are, of course, kept constant by feedback control but the total steroid secretion by the adrenal cortex changes dramatically with the changes in Δ^4-5α-reductase activity. The enzyme does not appear to be specific for corticosterone[39] but its physiological role with other steroids is unknown.

III. ALDOSTERONE

Aldosterone, the potent mineralocorticoid, is in large part metabolized by the liver (splanchnic system) with tetrahydroaldosterone being the major metabolite (approximately 35% of an injected dose for normal individuals).[40] A greater fraction of aldosterone is excreted with no alterations in ring A than is found for cortisol; however, while unmetabolized cortisol is excreted unconjugated, the aldosterone is excreted as the 18-glucuronide conjugate.[41] It is of interest that in situations such as pregnancy where the splanchnic extraction of aldosterone decreases, the extent of formation of tetrahydroaldosterone decreases slightly, although not significantly, and the excretion of the 18-oxoconjugate increases.[42] The 18-oxoconjugate is largely formed in extrahepatic tissues, in particular, the kidney.[41] Estrogen administration which also decreases extrasplanchnic extraction actually decreases the extent of formation of tetrahydroaldosterone; it is suggested that the lack of a significant decrease in pregnancy could be due to a contribution to aldosterone metabolism by the fetoplacental unit.[42] Liver cirrhosis leads to decreased formation of tetrahydroaldosterone and increased formation of the 18-oxoconjugate increases.[43] All these studies emphasize the importance of the liver in the Δ^4-reduction of aldosterone which is predominantly 5β; early studies on aldosterone metabolism by humans had indicated that most of the quantitatively important metabolites had the 5β configuration.[44,45] One of these reports also showed that tetrahydroaldosterone was the principal metabolite formed in vitro by the rat liver, indicating that Δ^4-5β-reduction also probably predominated in this species.[44] That aldosterone can be reduced to 5α-metabolites is supported by such metabolites being excreted in human urine;[46] however, all studies suggest that Δ^4-5α-reduction of aldosterone is of minor importance in the overall metabolism of this important mineralocorticoid. Of interest is the reported finding that 17-isoaldosterone is not metabolized to a tetrahydrometabolite, indicating that the liver Δ^4-5β-reductase system has specificity for the natural steroid configuration at position 17.[40]

In an excellent review of aldosterone metabolism in mild essential hypertension, Genest et al.[47] found an increase in 18-oxo-conjugate excretion, associated with a decreased aldosterone metabolic clearance rate and an increased plasma aldosterone level, suggesting a deficiency in liver aldosterone Δ^4-5β-reductase, the enzyme necessary for tetrahydroaldosterone formation. This enzyme may thus be very important in controlling the plasma level of aldosterone, the potent mineralocorticoid.

Another secretory product of the adrenal cortex is 18-hydroxydeoxycorticosterone. A major urine metabolite of this compound is 18-hydroxytetrahydrodeoxycorticosterone and its excretion can be used as an indicator of 18-hydroxydeoxycorticosterone.[48]

The 18-hydroxydeoxycorticosterone example is only one of several that could be discussed with regard to the adrenocortical secretory products. In most instances the tetrahydrometabolites can be used as a quantitative indicator of the level of production of the parent compound. The 3α-hydroxy-5βcompounds predominate in general as urine metabolites of the corticoids. Their formation results in compounds that can be conjugated for excretion. The interesting findings on the biological activity of 5α-dihydrocortisol and decreased Δ^4-5β-reduction of aldosterone in some cases of essential hypertension are examples of situations where abnormal steroid metabolism leads to clinical syndromes. In the first instance metabolism appears to stop after Δ^4-reduction while in the second instance there is a decreased Δ^4-reduction. Future studies will obviously be directed at further dissection of the problem at the tissue and enzymatic levels.

Studies[49,50] with synthetic glucocorticoids suggest that changes in the steroid structure which increase biologic potency when compared to the natural steroids may be in part due to their not being reduced by the liver Δ^4-reductases; this would lead to a reduced rate of clearance and an increased availability for exertion of a biologic effect.

IV. PROGESTERONE

Progesterone is an important intermediate in the biosynthesis of both adrenocortical and gonadal steroid hormones. It is itself an important hormone in female reproduction being secreted by the ovary in a reproducible pattern during the menstrual cycle. While the metabolism of a number of C_{19}-steroids yields ring A reduced products with significant and different biological activities, the evidence for such with progesterone, although it exists, is far less convincing. In general, it can be stated that the Δ^4-reduction of progesterone results in the formation of products which end up being reduced at the 3-ketone, giving 3-hydroxylated metabolites which in turn are conjugated with glucuronic acid and excreted, similar to the situation described for the corticoids; as stated previously, this will not be discussed in as much detail for progesterone. There is some recent evidence indicating that 5α-dihydroprogesterone may be formed in some progestin target tissues and then bind to the progestin receptors causing a biological effect. This is not as widespread a phenomenon as that which occurs for the androgens where the formation of 5α-dihydrotestosterone from testosterone is essential for the expression of androgenic activity in some target tissues.

One of the first urine steroid metabolites ever characterized was pregnanediol glucuronide.[51] Pregnanediol commonly refers to 5β-pregnane-3α,20α-diol, which is quantitatively the most important of the various possible reduced isomers. Urine pregnanediol levels have long been used as an indicator or indirect measure of progesterone production.[52] From a clinical point of view, this assay has in the past been very useful in the investigation of ovulatory disorders, i.e., the failure of a luteal phase peak of pregnanediol excretion suggests that ovulation and the formation of a corpus luteum have not occurred. This assay has been to a large extent replaced by the serum progesterone assay, which is a simpler assay to perform as well as being a more direct assay of progesterone production. The rise in the urine concentration of pregnanediol glucuronide parallels the rise in serum progesterone concentration following the time of predicted ovulation.[53] While pregnanediol is considered to be quantitatively the important metabolite, only 14% of the radioactivity is excreted as this metabolite following the administration of ^{14}C-progesterone to young men;[54] this percentage decreases significantly to 10% in older men. The significance of this latter finding is unclear. The formation of pregnanediol from progesterone is similar in nature to the formation of tetrahydrometabolites from the corticoids.

There have been a number of studies on the metabolism of progesterone in the uterus and its possible role in the action of progesterone. As progesterone is reduced primarily to 5α-metabolites in the rat uterus, the possibility was initially raised that the situation in this tissue might be similar to that for androgen target tissues such as the prostate whereby the active androgen intracellularly is 5α-dihydrotestosterone.[55] This seemed reasonable at first as the extent of Δ^4-5α-reduction was increased by estrogen administration which enhances the uterotrophic effect of progesterone. Conversely, prolactin administration was found to decrease Δ^4-5α-reductase activity and the extent of progesterone uterotrophic activity.[56] In a study[57] on the effect of estradiol administration on progesterone Δ^4-5α-reductase in the rat uterus, it was found that while enzyme activity was found in all tissue subcellular fractions, only that associated with the nuclei was increased by estrogen injections. Results similar to those just described for the rat have also been found for the rabbit, i.e., estrogen administration increases the ability of the uterus to form 5α-reduced products from progesterone in vitro;[58] recent in vivo studies indicated that estradiol administration caused no change in uterine progesterone Δ^4-reductase activity in ovariectomized rabbits.[59] It is of note that the rabbit uterine Δ^4-5α-reductase activity was greater for endometrium than for myometrium and was located primarily in the cell nuclei. However, later reports do not support the involvement of 5α-reduction in the expression of progesterone actions in the uterus. These include the finding that 5α-pregnane-3,20-dione has no effect on the rabbit uterus.[60] It was also found not to accumulate in the rat uterus following the administration of progesterone; this is apparently due to its rapid reduction to 5α-pregnane-3α-ol-20-one.[57] Recently it has been shown that 5α-pregnane-3,20-dione associates with the guinea pig cytosol progesterone receptor at a rate similar to that of progesterone itself;[61] however, it dissociates more rapidly than progesterone from the receptor. The lack of biologic activity of 5α-pregnane-3,20-dione in the uterus is probably explained by a combination of its rapid metabolism and its rapid dissociation from the cytosol progestin receptor. Such a conclusion is also supported by the finding that progesterone, but not 5α-dihydroprogesterone, induced the synthesis of uteroglobin (a uterine-specific protein in the rabbit) in ovariectomized rabbits.[58] This latter study also indicated that in the rabbit uterus 5α-dihydroprogesterone was not retained and was rapidly metabolized while progesterone was retained. It would appear that progesterone Δ^4-5α-reduction is not essential to the expression of progestational activity in the uterus. The enhanced progestational effects that occur in the estrogen-primed animal could be due to the reported increases in progesterone receptors which are used as markers for estrogen effects in some tissues.[62] Species differences in progesterone metabolism and effects must also be remembered as it has been reported[63] that while estrogen stimulated progesterone uptake by the guinea pig uterus, there is little metabolism of the progesterone; this contrasts with the findings in the rat. While Δ^4-5α-reductase activity for progesterone predominates in the uterus, there has also been a report of small amounts of 5β-dihydroprogesterone by human myometrium;[64] this finding has no known significance at this time other than it demonstrates that Δ^4-5β-reductase activity is not found exclusively in the liver.

While 5α-pregnane-3α,20-dione may be rapidly metabolized through the actions of 3α-hydroxysteroid dehydrogenase in the rat uterus and hence have no activity, this may not be true in other tissues. For example, this metabolite accumulated in the pituitary and hypothalamus when either ^{3}H-progesterone or ^{3}H-5α-pregnanedione was administered to rats;[65] this was also the primary metabolite in the tissue nuclei. This in vivo study followed an earlier in vitro study which indicated that 5α-pregnanedione accumulated along with precursor in the nuclei of rat medial basal hypothalamic tissue when slices of the tissue were incubated with labeled progesterone. The Δ^4-reduction of progesterone in the brain and the hypothalamus could theoretically modulate or mediate the effects of progesterone in either feedback control of gonadotropins or in behavioral effects. It was concluded[66] that Δ^4-5α-reduction of progesterone was probably not required for its effects on the induction of sexual receptivity in rats as the 5α-reduced metabolites were less active than progesterone itself when tested in estrogen-primed female rats. However, it should be noted that this conclusion did not take into consideration possible differences in the clearance rates for the different steroids. In the analogous situation in rats, 5α-dihydrotestosterone is cleared much more rapidly than is testosterone.[67,68] In ovariectomized rabbits 5α-dihydroprogesterone is cleared more rapidly than progesterone.[59] The Δ^4-5α-reductase activity in the rat medial basal hypothalamus is found primarily in the 1000 × g cell debris-membrane fraction but not in the nuclear fraction;[69] in terms of substrate specificity, 20α-hydroxyprogesterone was favored over progesterone. Estradiol was a noncompetitive inhibitor with a K_i approximately 100 times higher than the progesterone K_m. Progesterone Δ^4-5α-reductase activity is found in several areas of the rat brain as well as in the pineal gland.[70] The exact physiological role of the brain and hypothalamic progesterone Δ^4-5α-reductases requires further investigation.

Little et al.[71] reviewed in vivo progesterone metabolism and some of the reported findings are interesting in relation to the aspects of progesterone Δ^4-5α-metabolism just covered. The extrasplanchnic clearance of progesterone is more important than is that for some of the other steroids, in particular, testosterone. In the human, the extrasplanchnic clearance of progesterone amounts to 60% of the total; the corresponding figure for sheep is 73%. This means that metabolism in tissues other than liver is of major importance. Little et al.[71] looked in particular at brain, head, and uterine clearance and metabolism. These parameters were studied in the rhesus monkey and the rabbit. There was significant extraction of infused labeled progesterone and conversion to 5α-dihydroprogesterone by the head in both species. Conversions were higher in the rabbit. All areas of the brain had labeled progesterone but higher concentrations existed in the areas of potential action. In the uterus (studied in the rabbit only) 5α-dihydroprogesterne was also formed but was rapidly metabolized to other products and appropriate experiments lead to the conclusion that in this tissue Δ^4-5α-reduction was rate limiting in progesterone metabolism. Also, the rapid metabolism of 5α-dihydroprogesterone in the uterus suggested that it was not protected by binding to receptors and hence does not play a significant role in progesterone actions, at least in the rabbit.

That the skin can actively reduce progesterone is evident from several reports. Mauvais-Jarvis et al.[72] studied the in vivo metabolism of progesterone in humans. When administered intravenously, eight times more 5β-pregnane-3α,20α-diol than 5α-pregnane-3α,20α-diol was formed. When administered percutaneously (rubbed on the skin) more 5α-pregnane-3α,20-diol than 5β-pregnane-3α,20α-diol was formed. Hence it was concluded that progesterone underwant Δ^4-5α-reduction in the skin. This finding was supported by the in vitro studies of Gomez et al.[73] who found that the enzyme activity was associated with the microsomes. It is possible that both progesterone and testosterone are reduced by the same Δ^4-5α-reductase.[73] It is of interest that in the in vivo studies progesterone, when administred intravenously, was metabolized similarly for men and women and for men receiving estrogen.[74] This contasts with the findings for testosterone metabolism where differences in the level of sex hormone binding globulin influence the extent of Δ^4-5α-reduction;[75] changes in the level of transcortin which binds progesterone do not have the same influence on its metabolism.

Progesterone may have some influence on the extent of Δ^4-5α-reduction through changes in the rate of its own secretion. In measurements of plasma progesterone and 5α-pregnane-3,20-dione throughout the normal menstrual cycle, the levels of the latter are slightly higher during the follicular phase, but during the luteal phase are much less than those of progesterone which is being secreted at a much greater rate;[76] the ratio of progesterone to 5α-pregnane-3α,20-dione is approximately 7:1 during the luteal phase. Human endometrial progesterone Δ^4-5α-reuctase which is found in the microsomes changes during the menstrual cycle with the levels being significantly higher during the early proliferative phase.[77] In the same study the levels were found to increase in endometrial carcinoma tissues as the tumor became less differentiated. Also 5β-dihydroprogesterone was formed in the endometrial-soluble fraction. In some studies on Δ^4-5α-reduction during pregnancy the metabolism of progesterone by the human feto-placental membranes was examined. Both the chorion and amnion had Δ^4-5α-reductase activity with the latter having the highest levels.[78] The enzyme levels were found to decrease during pregnancy[78] with the level being almost nil in the chorion at term. This raised the possibility that progesterone exerts its actions via its metabolites and that a decrease in metabolism initiated labor or that the level of enzyme may be related only to progesterone availability. The fetal membranes have the ability to bind 5α-pregnane-3,20-dione.

A number of other tissues have the ability to form progesterone Δ^4-reduced metabolites and the effects of 5α-dihydroprogesterone have been examined in several biological systems. For example, 5α-pregnane-3,20-dione caused ovulation in PMSG-treated immature rats although it was less potent than progesterone.[79] Progesterone in vivo has not activity by itself on preputial glands but it does alter the in vitro metabolism of labeled progesterone by the tissue possibly by increasing both Δ^4-5α-reductase and 20α-hydroxysteroid dehydrogenase activity;[80] it was suggested that through these changes progesterone may regulate its own levels within the tissue. Progesterone is synergistic with other hormones in controlling mammary growth and its metabolism by the rabbit mammary gland has been studied;[81] both 5α-metabolites were identified indicating another tissue where Δ^4-5α-reduction can occur. Again, the relationship of progesterone metabolism to its actions in mammary tissue is unclear. Mouse vaginal tissue actively forms 5α-reduced metabolites of progesterone; the progesterone Δ^4-5α-reductase activity in this tissue does not appear to be affected by estrogen administration but the vivo and in vitro experiments suggest that it is the rate limiting step in the formation of products redced at C-3 or C-20.[82]

A system that has been extensively used as a model system for studying progesterone action is the chick oviduct. A number of 5α-reduced progesterone metabolites will displace progestrone from its receptors in this system.[83] When progesterone metabolism was studied in different areas of the chick oviduct, Δ^4-5α-reduction was found in th shell gland and comb but not in the magnum, the site of avidin synthesis.[84] In the shell gland 5α-dihydroprogesterone was found associated with the chromatin. The Δ^4-5α-reductase which reduced both progesterone and testosterone was also found associated with the chromatin. The actual significance of the 5α-reduction in the shell gland remains to be eluciated.

A number of situations have been described where there is Δ^4-reduction of progesterone at or near sites of its actions. The precise role remains to be elucidated in most instances.

V. C_{19}-STEROID Δ^4-REDUCTASES

The Δ^4-reduction of C_{19}-steroids has been studied in great detail. This is due at least in part to the 5α-dihydro steroids having androgenic activity with 5α-dihydrotestosterone being a more potent androgen in bioassays than testosterone itself.[85] The formation of 5α-dihydrotesosterone is essential to the expression of testosterone activity in some tissues.[55] For this reason, the role of Δ^4-reduction in C_{19}-steroid metabolism will be discussed in greater detail than it was for the corticoids and progesterone.

The principal C_{19}-steroids secreted by the adrenal cortex and gonads include dehydroisoandrosterone both in the free for and as the 3-sulfate, Δ^4-androstenedione, 11β-hydroxy-Δ^4-androstenedione, and testosterone. While dehydroisoandrosterone and dehydroisoandrosterone sulfate are excreted in part in the form of the latter, a major fraction of these two steroids as well as practically all of the others are excreted as metabolites where the ring A double bond has been reduced. These metabolites fall into the group of urine compounds called 17-ketosteroids and androstanediols (see Figure 3). Dehydroisoandrosterone and its sulfate are metabolized in part by way of the Δ^4-3-ketosteroid structure (Δ^4-androstenedione) and Δ^4-reductases to compounds such as androsterone and etiocholanolone[86] prior to excretion. As an indication of the extent of metabolism of testosterone in humans, it is known that approximately 6 mg of testosterone is produced per day by an adult male;[87] however, only 100 μg is excreted with the Δ^4-3-ketosteroid structure intact.[88] This is in the form of testosterone glucuronide.

Many studies on C_{19}-steroid Δ^4-reductases and Δ-^{4}-reduction have been performed in humans and rats. As for the glucocorticoids and progesterone, the studies in humans can in many instances be considered indirect in that they deal with in vivo studies whereby labeled steroids are injected intravenously and the nature and relative

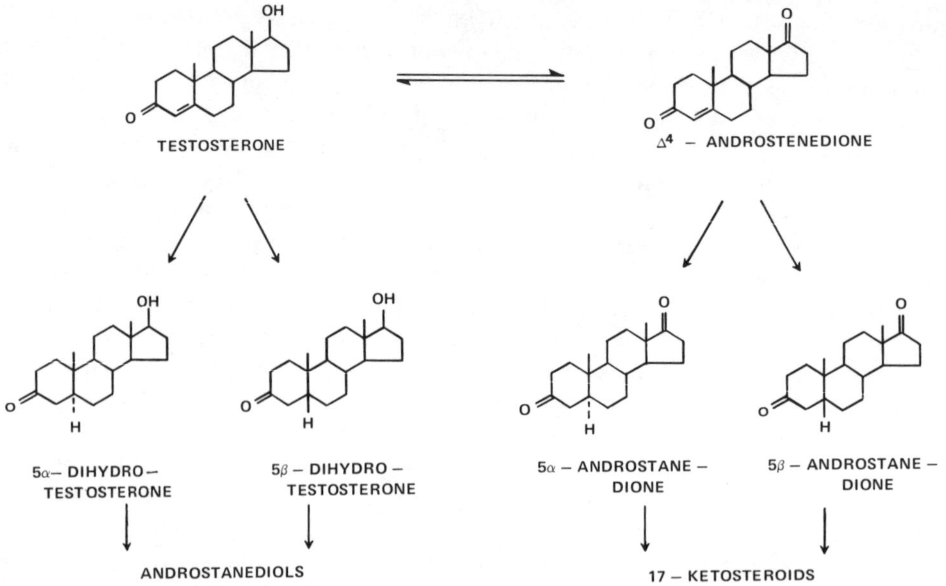

TESTOSTERONE

Δ^4 – ANDROSTENEDIONE

5α– DIHYDRO– TESTOSTERONE

5β – DIHYDRO – TESTOSTERONE

5α – ANDROSTANE – DIONE

5β – ANDROSTANE – DIONE

ANDROSTANEDIOLS

17 – KETOSTEROIDS

FIGURE 3. Pathways of androgen metabolism involving Δ^4-reduction.

amounts of the ring A saturated steroids in the urine are examined. Conclusions are then drawn with regard to the relative levels of Δ^4-reductase activity in different experimental situations or to the relative levels of Δ^4-5α- and Δ^4-5β-reductase activities in liver. Such conclusions are, in general, reasonable since the liver is quantitatively the principal site of C_{19}-steroid clearance. The studies in rats have been both in vivo and in vitro in nature. The Δ^4-reductase activity has been characterized for several rat tissues and the findings complement in vivo results. Some in vitro studies have been performed in dog prostate because the dog is the only animal species that has been found to get benign prostatic hypertrophy and hence it is used as a model for the human condition.

In terms of androgenic activity, testosterone is the important C_{19}-steroid in the circulation. The clearance of testosterone from the circulation when measured in terms of the metabolic clearance rate correlates with the unbound testosterone level.[89] The current general concept is that bound testosterone is not free to enter tissues where it can exert a biological effect or be irreversibly metabolized and hence cleared from the circulation. As noted earlier, it is believed that Δ^4-reduction is the rate limiting step in steroid metabolism, particularly in the liver. The rate of clearance, although controlled in large part by binding to sex hormone binding globulin, can be increased by increasing the liver Δ^4-5α-reductase levels. This occurs with administration of the progestin medroxyprogesterone acetate.[90] This compound increases the the metabolic clearance rate of testosterone in men.[91]

Because Δ^4-reduction is probably rate limiting in testosterone metabolism, relatively little 5α-dihydrotestosterone (the product of Δ^4-5α-reduction of testosterone) leaves the liver; it is metabolized further to 17-ketosteroid- and androstanediol-type metabolites. This will account at least in part for the low levels of 5α-dihydrotestosterone in plasma relative to testosterone. Also, 5α-dihydrotestosterone is cleared from the circulation even more slowly than testosterone.[92] This can be readily explained by the stronger binding of 5α-dihydrotestosterone to sex hormone binding globulin.[93] We have measured the metabolic clearance rate of testosterone and 5α-dihydrotestosterone in adult rats, a species that has no specific binding protein for androgens in plasma.[94] In this instance,[67,68] the MCR for testosterone (18ℓ/24 hr/100 g body weight) is significantly lower than that of 5α-dihydrotestosterone (87 ℓ/24 hr/100 g body weight); this finding supports the concept that Δ^4-reduction is rate limiting in testosterone metabolism within tissues.

As will be detailed, many androgen target tissues have the capacity to convert testosterone to 5α-dihydrotestosterone. This metabolite, which in some bioassays is more potent than testosterone itself, is probably the important intracellular andogen in these target tissues. The presence of the target tissue Δ^4-5α-reductases accounts for a higher urine androsterone/etiocholanolone ratio (for the ^{14}C metabolites of ^{14}C-testosterone or for endogenous precursors) for men than for women.[95] Approximately 40% of a dose of intravenously administered ^{14}C-testosterone is converted to urine androsterone and etiocholanolone.[95] The mean ^{14}C-androsterone/^{14}C-etiocholanolone ratios were 1.5 and 0.9 for young men and women, respectively, and 1.0 and 0.4 for old men and women respectively. The ratios were significantly different for both age groups. In support of the male androgen target tissues playing a significant role in the production of the 5α-metabolite androsterone was the finding that the ratios were not significantly different for boys and girls (1.7 and 1.9, respectively). There was no sex difference for the 5α/5β androstanediols for any of the age groups. That the sex hormone binding globulin influences the entry of testosterone into target tissues where Δ^4-5α-reduction occurs is supported by the finding of less 5α-metabolites in situations where the level of the binding protein has been increased by estrogen administration.[96]

In the liver the evidence in general supports the presence of soluble Δ^4-5β-reductase(s) and microsomal Δ^4-5α-reductase(s) for C_{19}-steroids. The early studies of Tomkin's laboratory indicated the presence of a series of each of the stereospecific enzymes with the members of each series differing in substrate specificity.[14] This included more than one enzyme in each subcellular compartment for C_{19}-steroids. Evidence has abso been reported for more than one soluble Δ^4-5β-reductase in male pig liver;[97] in this instance evidence was found for at least two Δ^4-5β-reductases, one favoring testosterone as substrate and the other cortisol.

In most situations where liver testosterone Δ^4-reductase is altered by experimental manipulations, changes in activity levels appear to occur in the microsomal Δ^4 5α-rcductase levels. As mentioned previously, the levels of hepatic Δ^4-5α-reductase in humans are increased by oral administration of the synthetic progestin medroxyprogesterone acetate.[90] These significant increases in enzyme activity result in an increase in the metabolic clearance rate of testosterone;[91] this again supports the theory that Δ^4-reduction is rate limiting in the overall clearance of testosterone from the circulation. Experiments in rats also support these findings; the prostates of male animals receiving medroxyprogesterone acetate had less of a response to testosterone injections than did normal animals. Again, the liver microsomal Δ^4-5α-reductase activity was higher than that found for normal animals.[91]

Estrogen administration also increases liver microsomal testosterone Δ^4-5α-reductase in male rats[98] and this again results, at least in part, in an increased rate of clearance of testosterone from the circulation.[67] As mentioned previously in the section on corticoids, the effects in female rats are less clear with there being reported decreases as well as no change in activity. The situation in humans is more complex as estrogen administration causes an increase in the levels of sex hormone binding globulin[93] which has a major influence on the metabolic clearance rate of testosterone, causing it to decrease very significantly.[99]

Another hormone which causes an increase in liverΔ^4-5α-reduction in male rats is thyroxine.[100] Again, there may be a decrease in the levels of the enzyme activity in the female liver. Although not assayed directly in human liver samples, thyroxine would appear to also increase microsomal Δ^4-5α-reductase levels as the ratio of urine 5α-to 5β-metabolites (androsterone/etiocholanolone) of testosterone increases.[101] Again, there is probably not a great physiological significance of this in humans when there are changes in thyroxine status because thyroxine also increases the levels of sex hormone binding globulin[93] which will prevent testosterone or the more potent 5α-reduced androgens produced in the liver from entering target tissues.

An interesting situation where there is an apparent decrease of deficiency in liver Δ^4-5α-reductase levels is that of acute intermittent porphyria. There has been considerable evidence to suggest that 5β-reduced steroids induce ALA synthesis in liver. This particular enzyme catalyzes the rate limiting step in porphyrin biosynthesis. While there have been one or two reports suggesting that 5α-reduced steroids are just as effective as the 5β-isomers at inducing ALA synthetase in in vitro systems,[102] the facts indicate that in patients with acute intermittent porphyria testosterone is metabolized to a much greater extent to 5β-metabolites than is found for normal subjects.[103] This situation can be exacerbated by the administration of phenobarbital which appears to increase microsomal drug-oxidizing enzymes while at the same time decreasing Δ^4-5α-reductase acitivty.[104] The ratio of 5β/5α-testosterone metabolites can be decreased in patients with acute intermittent porphyria by the administration of triiodothyronine.[103] Hence, the liver apparently has the capacity to have its Δ^4-5α-reductase levels increased by appropriate stimuli, but in normal situations the levels are low. The administration of the thyroid hormone caused the patients to become hypermetabolic. The altered steroid metabolism is not restricted to testosterone. Some 80% 11β-hydroxy-Δ^4-androstenedione is normally metabolized in vivo by humans to 5α-reduced metabolites. In acute intermittent porphyria only 35% of administered labeled 11β-hydroxy-Δ^4-androstenedione appears in the urine as 5α-metabolites.[103]

The specificity of the effect on ALA synthetase of 5β-steroid metabolites is not entirely clear at this time. While earlier in vitro test systems, e.g., chick blastoderm cultures, indicated that 5β-steroids were more effective as inducers than the 5α-steroids, they contained serum in the incubation medium.[103] More recent studies[102] with rat liver cultures suggest that 5β- and 5α-steroids are equally potent. These findings could possibly be due to a greater extent of binding of the 5α-metabolites by serum proteins which would prevent their entry into the cells. This possibility awaits further investigation.

The sex differences in rat liver Δ^4-5α-reductases (higher activity in females than males) may be programmed very early in life by androgens operating through pituitary factors.[105] This concept was supported by the finding of pituitary androgen receptors in the male neonatal rat but not in the female animal.[106] This appears to be species specific as programming in other species appears to occur later in life.

As indicated above, liver Δ^4-reductases are important in the quantitative clearance of C_{19}-steroids from the circulation. However, it must be reemphasized that Δ^4-5α-reduction of the steroids is of tremendous physiological significance in the expression of androgenic activity in some, but not all, tissues where effects are exerted. As will be discussed in greater detail below, Δ^4-5α-reduction is of major importance in the expression of androgenic activity in tissues of the male reproductive tract, i.e., testes, epididymis, prostate and seminal vesicles, hair follicles, skin, and submaxillary and preputial glands. In fetal development and differentiation Δ^4-5α-reduction is required for formation of the prostate and external genitalia but is not involved in the androgenic control of differentiation and formation of internal genitalia. During maturation Δ^4-5α-reduction of testosterone by the testes is of more importance in young animals than in older animals. Δ^4-5α-Reduction is probably not involved in the actions of androgens in tissues such as the muscle and the kidney. The involvement of Δ^4-5α-reduction in the effects of androgens in actions at the level of the brain, adenohypophysis, and hypothalamus is far from being clear.

From a historical point of view, Δ^4-5α-reduction was first determined to be important in the expression of the androgenic activity of testosterone for the prostate. Pearlman and Pearlman[107] had reported in 1961 that when [^{14}C]-Δ^4-androstenedione was infused into adult male rats, the radioactivity in the prostate was principally in the form of 5α-metabolites. In 1968, the classic studies of Bruchovsky and Wilson[108] demonstrated that when ^{3}H-testosterone was injected into functionally hepatectomized-nephrectomized adult rats quantitatively the important labeled metabolite found in the prostate was 5α-dihydrotestosterone. When the prostate was fractionated into its subcellular components and the ^{3}H-metabolites examined in each component, practically all the radioactivity in the nucleus was in the form of 5α-dihydrotestosterone. Reports from the laboratories of Anderson and Liao[109] and Baulieu et al.[110] in 1968 also showed that labeled 5α-dihydrotestosterone was the major compound found in the prostate following the administration of ^{3}H-testosterone to rats. These studies coupled with earlier knowledge that 5α-dihydrotestosterone was a more potent androgen than testosterone itself stimulated many studies on the relationships of androgen metabolism to the expression of biologic activity in the prostate and other androgen-dependent target tissues. These include reports on the specificity of the binding of compounds by the androgen receptor and the nature and control of Δ^4-5α-reductase activity.

In general, it is now believed that while testosterone is quantitatively the important androgen in the circulation, 5α-dihydrotestosterone is the important androgen intracellularly in the prostate. It has been shown[111] that the androgenic activity of a number of natural C_{19}-steroids correlated well with the amount of labeled 5α-dihydrotestosterone which appeared in the rat prostate following administration of these steroids containing tritium. In a series of 11 different species it was found[112] that the rate of conversion of testosterone to 5α-dihydrotestosterone correlated with the ultimate size of the gland; dog and man had high conversion rates. The soluble androgen receptor in the prostate which binds steroid enters the nucleus, binds to chromatin acceptor sites to stimulate transcription leading to expression of androgenic activity, and has a higher affinity for 5α-dihydrotestosterone than for any other androgen.

The basic characteristics of the Δ^4-5α-reductase enzyme activity have been studied largely in the rat although measurements of activity in different situations have certainly been made in other species including man. The enzyme activity is located predominantly in the microsomes and nuclei with approximately half being found in each subcellular site.[113] In tissues such as the liver where steroid metabolism is probably not involved in actions of the hormone, significant levels of activity of enzymes involved in the metabolism are not associated with the nucleus. Also, it appears[113] that there is not a series of Δ^4-5α-reductases in the prostate, at least in the nucleus, differing in substrate specificity, as there is in the liver. The reduction of any of the steroids which are converted to 5α-metabolites by the prostate nuclei appears to be performed by one enzyme. The nuclear enzyme can be extracted by high salt concentrations. It appears[114] to be on the outer nuclear membrane as procedures such as washing with Triton® X-100 which remove this membrane without destroying the nucleus removes enzyme activity. The enzyme is strongly inhibited by some divalent cations such as Hg^{++} and Zn^{++}.[113] This is of interest since the prostate, in particular the dorsolateral prostate, accumulates Zn^{++} under the influence of androgens.[115] Hence, Zn^{++} has a potential role in controlling the rate of Δ^4-5α-reduction of testosterone. The exact physiological significance of the nuclear Δ^4-5α-reductase remains an important question as one could postulate that testosterone is reduced to 5α-dihydrotestosterone prior to its binding to the cytosol receptor which then enters the nucleus, and hence the need for the nuclear enzyme is left in doubt. One might suggest that it prevents the entry of receptor-testosterone complexes and hence ensures that only 5α-dihydrotestosterone enters the nucleus. This would mean that the nuclear enzyme is an insurance mechanism. Rat prostatic Δ^4-5α-reductase is under hormonal control in that its specific activity is decreased by castration, adrenalectomy, and estrogen administration while it is increased by testosterone administration.[116]

There have been a number of studies on androgen metabolism and levels in human prostate disease situations. Dihydrotestosterone contents for hypertrophic prostate (0.60 μg/100 g) were higher than those for normal prostate (0.13 μg/100 g); however, the rate of conversion of testosterone to 5α-dihydrotestosterone was similar for both types of tissue.[117] In this first report,[117] the 5α-dihydrotestosterone content was higher in the periurethral area of the gland when compared to the outer regions; prostatic hypertrophy begins in the periurethral area. 5α-Dihydrotestosterone administration to dogs did not consistently lead to prostatic hyperplasia; its further reduction product 5α-androstane-3↓,17β-diol did.[118] This latter finding is interesting in view of the fact that prostatic androgen receptors bind 5α-dihydrotestosterone much more avidly than the androstanediol.

Recently, Ishimaru et al.[119] have reported that in elderly men with benign prostatic hypertorphy the plasma levels of testosterone are lower than those for young men; the levels of 5α-androstane-3α,17β-diol are similar for the two groups but those of 5α-dihydrotesterone are higher. Kinetic studies indicated that the production rate for testosterone was decreased for the patients but that for 5α-dihydrotestosterone was similar for the two groups of subjects. In young men 80% of the circulating 5α-dihydrotestosterone was derived peripherally from testosterone while in the older men 50% was secreted directly, possibly by the testes or, more probably, the prostate. Evidence that the prostate secretes 5α-dihydrotestosterone has been reported.[120] Vermeulen and De Sy[121] have found the plasma testosterone:5α-dihydrotestosterone ratio is lower for men with benign prostatic hypertrophy than for normal old men; this is due to a relatively higher 5α-dihydrotestosterone level than that of testosterone. Most of the human prostatic testosterone Δ^4-5α-reductase has been localized in the stroma with relatively much less activity being associated with the epithelium.[122] Habib et al.[123] have found a much higher tissue 5α-dihydrotestosterone:testosterone ratio for benign prostatic hypertrophic tissue as compared to prostatic cancer tissue. The androgen receptor in human benign prostatic hypertrophic tissue has been studied by a number of investigators. While there have been some problems with some types of methodology in distinguishing receptor from plasma sex hormone binding globulin binding, current evidence indicates that the receptor is quite specific for 5α-dihydrotestosterone and endogenousy 90% of occupied receptor is bound to this steroid. Giorgi et al.[124] used a perfusion technique with slices of normal and hyperplastic human prostatic tissue and studied the uptake and metabolism of testosterone, Δ^4-androstenedione, and 5α-dihydrotestosterone. Testosterone was converted to 5α-dihydrotestosterone at a significant rate and this latter compound was retained to significantly greater extent than either of the other two androgens. Tissue uptake for hyperplastic tissue was greater than for normal tissue.

While there is considerable evidence to suggest an association between benign prostatic hyperplasia and 5α-dihydrotestosterone, the aberration is still unclear; Δ^4-5α-reduction appears to be normal. A question that has not been extensively examined is whether there is less metabolism of 5α-dihydrotestosterone to other androgen metabolites.

The relationship between androgens, androgen metabolism, and prostatic carcinoma is less clear. It has been widely held that many, but certainly not all, prostatic carcinomas are androgen dependent. As mentioned above, 5α-dihydrotestosterone levels are reported to be higher in benign prostatic hyperplastic tissue than in prostatic cancer tissue. Geller et al.[125] have also found a lower average tissue 5α-dihydrotestosterone concentration for cancer tissue but the range of values was wide and ⅔ of the values overlapped. Treated cancer patients showed a wide range of values including several with estrogen administration with very high values. Farnsworth and Brown[126], on the other hand, found no differences in the relative amounts of testosterone, 5α-dihydrotestosterone, and 5α-androstane-3α,17β-diol when concentrations in benign prostatic hyperplastic or prostatic cancer tissues were compared. These findings are difficult to explain in view of the fact that, in general, increased 5α-dihydrotestosterone is associated with prostatic hyperplasia.

Estrogen is often used as a therapy for patients with prostatic cancer. It is believed that this therapy decreases testosterone production by inhibition of LH production[127] or possibly by direct effects on the testes.[128] However a number of studies have sought possible direct effects on the prostate including the effects on the Δ^4-5α-reduction of testosterone. Very high levels of estrogen will inhibit rat prostatic Δ^4-5α-reductase activity in vitro but the levels required for this effect are probably never attained in vivo, even with pharmacological doses.[129] Giorgi's laboratory[124] has utilized the slice perfusion technique to study the effects of estradiol on androgen uptake and metabolism by normal, hyperplastic, and carcinomatous human tissue. The results were varied: at lower estradiol concentrations, clearance of the androgens decreased for the hyperplastic tissues and increased for the other two while at higher estradiol concentrations the entry and clearance decreased for all tissues. The conversion of testosterone to 5α-dihydrotestosterone varied with entry of the precursor into the tissue. The higher retention of the reduced product in the tissue may therefore be related to high affinity receptor concentration rather than enzyme activity. The conversion was noted to be low in several samples of cancer tissue, again indicating a wide spectrum of Δ^4-5α-reductase activities in prostatic cancer tissues. In organ cultures of human prostatic hyperplastic and prostatic carcinoma tissue, estradiol decreased the production of dihydrotestosterone from both testosterone and Δ^4-androstenedione and at low concentrations increased the production of androstanediols. In general, the level of Δ^4-5α-reduction was lower for cancer tissue than for hyperplastic tissue. Prout et al.[130] also found that neoplastic prostatic tissue converted less testosterone to 5α-dihydrotestosterone than did either normal or hyperplastic prostatic tissue. In their neoplastic samples it was found that prostatic stroma accounted for a very significant fraction of the Δ^4-5α-reductase activity. In a study on the possibility of using Δ^4-5α-reductase in cultured prostate tissue as a target enzyme for antiprostatic drugs the results with estrogen-type compounds and human cancer issue were variable.[131]

In the prostates of the human and the dog there is certainly good evidence that 5α-dihydrotestosterone is the active intracellular androgen and there is an association between increased levels of this hormone and benign prostatic hyperplasia. It remains to be determined as to whether the increased concentrations are due to increased formation through, for example, either increased Δ^4-5α-reductase activity, decreased 3α-hydroxysteroid dehydrogenase, or 17β-hydroxysteroid dehydrogenase activities, or due to increased receptor levels which retains the 5α-dihydrotestosterone in the tissue. There, however, does not appear to be an association between increased levels of this androgen and prostatic carcinoma. In fact, it would appear that as the neoplasm becomes less differentiated there is a loss of Δ^4-5α-reductase activity. Such a conclusion is also supported by an animal model where not only was decreasing Δ^4-5α-reductase activity reported for the aging rat but also for adenocarcinomatous tissues with time.[132]

Other parts of the male reproductive system including the testes all reduce testosterone to 5α-reduced metabolites. The seminiferous tubules are an androgen target tissue since androgens are required in the development and maintenance of mature sperm.[132] The metabolism of testosterone by the individual components of the testes (the interstitial tissue and seminiferous tubules) has been studied for both humans and rats.[133,134] The findings are not dissimilar. The seminiferous tubules have Δ^4-5α-reductase activity which results in the formation of 5α-dihydrotestosterone and 3α- and 3β-androstanediols. Interstitial tissue primarily converts testosterone to Δ^4-androstenedione. For human testes, the levels of Δ^4-5α-reductase varied with the type of testicular tissue, suggesting some role in the control of expression of androgenic activity, although some differences in measured activity could be due to differences in endogenous testosterone concentrations. For example, cryptorchid testes from two adolescents with poor development of the germinal epithelium had low Δ^4-5α-reductase activity. Within the seminiferous tubules, the Sertoli cells appear to be the site for Δ^4-5α-reduction;[134] such conclusions arise from several experimental situations including ones where the Sertoli cells have been separated from other cell types or other cell types have been in large part destroyed by in vivo treatment. Injections of 5α-dihydrotestosterone to adult rats decrease circulating gonadotropin levels (testicular testosterone levels), but increase testicular androgen binding protein levels;[135] this latter protein is a secretory product of the Sertoli cell. This finding indicates that this reduced testosterone metabolite which is formed in the Sertoli cells can maintain cell function, and hence supports the role of Δ^4-5α-reductase in the expression of testosterone actions. The finding that exogenous 5α-dihydrotestosterone was less effective than testosterone in maintaining spermatogenesis in hypophysectomized rats was explained by the more rapid clearance of the reduced steroid which led to lower plasma and rete testes fluid androgen levels;[136] hence, less androgen was available to exert an effect than when testosterone was administered.

The ability of the rat testes to form 5α-reduced metabolites of testosterone changes during development. The prepubertal rat testes form much greater quantities of 5α-androstane-3α,17β-diol from progesterone than do the newborn or adult testes.[137] Similar findings were also reported using testosterone as a substrate.[138] Again, species differences appear to exist as the prepubertal rabbit testes produces primarily testosterone associated with only minor amounts of 5α-reduced androgens.[139]

The rat epididymis also has testosterone Δ^4-5α-reductase activity.[140] Its characteristics were similar to those of the prostatic enzyme in that it was found in both the nuclei and microsomes, it was strongly inhibited by Zn + +, and progesterone and epitestosterone were potent competitive inhibitors. The nuclear Δ^4-5α-reductase activity level was found to be fivefold higher in the caput corpus than in the cauda segment of the epididymis;[141] it decreased to undetectable levels following castration and was only partially returned to normal by testosterone administration. That hormones in the rete testes fluid other than testosterone may also be involved in maintenance of epididymal Δ^4-5α-reductase activity was suggested by findings following ligature of the efferent ducts[142] which caused a significant decrease in enzyme activity in the caput but not other segments of the tissue.

The brain, including the hypothalamus and adenohypophysis, has been the subject of some studies with regard to testosterone metabolism. Testosterone is involved in gonadotrophin feedback control and male behavioral patterns. Hence, the question is raised as to whether Δ^4-5α-reduction is involved in these actions of testosterone as it is in target tissues such as the prostate. Some actions of testosterone in the brain may involve aromatization of testosterone to estradiol rather than Δ^4-5α-reduction.[143]

There appears to be some species differences in the extent of Δ^4-5α-reduction by the brain and pituitary. Testosterone Δ^4-5α-reduction was demonstrated in all areas of the human fetal brain and to a lesser extent in the pituitary.[144] The levels of reduction were less for adult tissues raising the possibility that this steroid conversion is important during sexual development. Estrogen formation occurred in some brain areas but not in the pituitary. In the adult rat, testosterone Δ^4-5α-reduction was greater in the pituitary than in the brain.[145] Whether there is a decrease in brain Δ^4-5α-reduction with age in the rat is not as clear as the studies in the newborn[146] and adult rats[147] were perfotmed by different laboratories. Another study in adult rats led to the suggestion that 5α-dihydrotestosterone may be involved in gonadotropin control as following infusion of labeled testosterone this metabolite predominated in pituitary nuclei.[148] Labeled estradiol predominated in limbic system cell nuclei. Interestingly, 5α-dihydrotestosterone was lowest in the midbrain, the site of highest Δ^4-5α-reductase activity. The microsomal testosterone Δ^4-5α-reductase activity from the male rat pituitary was increased following castration;[149] testosterone administration maintained normal enzyme activity. 5α-Reduced androgens were less effective than testosterone.

The ability of a number of testosterone metabolites to decrease LH and FSH levels in castrated adult rates was studied by Verjans and Eik-Nes.[150] The effects were compared with the abilities of the same steroids to increase prostatic weight in the same animals. 5β-Metabolites had no effect on either parameter. 5α-Metabolites had varying activities with 5α-dihydrotestosterone and 3α-androstanediol being the most potent. Hence, the 5α-metabolites formed by Δ^4-5α-reductase in the brain and the pituitary may be playing a role in gonadotropin feedback control by testosterone, as 5α-reduced C_{19}-steroids cannot be aromatized and hence their effects cannot be exerted after conversion to estrogens as could hold for testosterone. It is possible that all the 5α-reduced compounds are active after conversion to 5α-dihydrotestosterone such as occurs in the prostate. In rats castrated prior to puberty, testosterone, 5α-dihydrotestosterone, and 5α-androstane-3α,17β-diol all suppressed gonadotropin levels although higher doses were required for FSH than for LH.[151] In this study testosterone had more of an effect on seminal vesicle weight than did the 5α-metabolites. Such was not the case for the ventral prostate leading to the conclusion that, at least in this system, slightly different mechanisms or metabolic transformations were operating in the different tissues.

With regard to androgenic control of sexual behavior, studies in castrated adult male rats[152] with dihydrotestosterone and 19-hydroxytestosterone (an estradiol precursor) suggested that both testosterone metabolites were required. The 19-hydroxytestosterone restored aspects of sexual behavior associated with brain functions while 5α-dihydrotestosterone appeared to be responsible for sperm production by the reproductive tract. Mounting behavior in adult rats was influenced by neonatal testosterone but not by 5α-dihydrotestosterone administration;[153] there was increased sensitivity to testosterone injections in adulthood in animals receiving the androgen neonatally. This again supports the concept that Δ^4-5α-reduction may not be involved in the effects of testosterone on sexual behavior.

That Δ^4-5α-reduction may not be required for control of gonadotropin production by the pituitary was suggested by a study with 17β-hydroxy-17α-methylandrosta-1:4-dien-3-one which is not significantly reduced by the prostate.[154] This synthetic steroid was as effective as testosterone at decreasing LH levels in castrated rats while it did not maintain ventral prostate weight.

The effects of androgens on skin and hair growth have been studied with regard to their role on sebaceous gland secretions and their possible relationship to hirsutism. Δ^4-5α-Reductase activity is present in human skin with activities being higher in the genital areas.[155] Sebaceous glands of animals have high Δ^4-5α-reductase activity. Human male skin from suprapubic areas has a higher testosterone Δ^4-5α-reductase activity than does female skin from the same areas.[156] The enzyme activity levels are low for children and increase in males at the time of puberty. Levels for hirsute women were in the normal male range. Again, as for male reproductive tract androgen target tissues, Δ^4-5α-reductase may also be involved in androgen actions.

An interesting study on the quantitation of testosterone and 5α-dihydrotestosterone in a number of male guinea pig tissues showed that the levels of 5α-dihydrotestosterone correlated with the capacity of the tissue to form it from testosterone and then retain it through the presence of receptors.[157] Levels were much higher in the prostate and seminal vesicles than in plasma. Interestingly, the levels in several types of muscle were very low.

Testosterone Δ^4-5α-reduction plays a critical role in sexual development. In the androgenically controlled development of the male reproductive tract of the fetus, formation of 5α-dihydrotestosterone is required for differentiation of the urogenital sinus into the prostate gland while testosterone itself is responsible for the differentiation of the Wolffian duct into the seminal vesicles and other internal structures. The involvement of Δ^4-5α-reduction in the expression of the androgenic activity of testosterone in the seminal vesicle appears to occur neonatally. Such conclusions are supported by studies on testosterone metabolism by human fetal tissues during sexual differentiation.[157]

In male pseudohermaphroditism there is not proper development of external genitalia, with a resultant ambiguous situation. There is not one abnormality in the expression of androgenic activity in this group of patients but rather they can be subdivided into small groups based on differences in biochemical deficiencies. It is now documented that a few of these genotypic males have a deficiency in Δ^4-5α-reductase. Two groups[158,159] originally described male psuedohermaphrodite patients where a number of studies on testosterone metabolism suggested a deficiency in testosterone Δ^4-5α-reduction. These included decreased Δ^4-5α-reductase activity in slices of skin from the affected individuals when compared to normal subjects and an increased 5β/5α ratio in testosterone metabolites excreted in the urine following administration of labeled metabolites. The skin, being an androgen target tissue, can be used as an indicator of androgen actions with regard to both Δ^4-5α-reductase and 5α-dihydrotestosterone binding by androgen receptors. In patients with Δ^4-5α-reductase deficiency, gonadotropin feedback is abnormal while sexual drive is normal; this again supports the concept that the enzyme may be required for androgen actions in the pituitary but not the brain. Studies to date suggest that binding can be normal even if reduction is deficient. Some individuals who are genotypically male yet are phenotypically female (e.g., testicular feminization syndrome) appear to have normal tissue Δ^4-5α-reductase levels but do not have androgen receptors for binding the 5α-dihydrotestosterone.[160] Rat and mouse models of this situation have been very useful in the study of this latter situation.[161] Because of the lack of the prostate and seminal vesicles in such animals, the kidney and submandibular glands have been studied as androgen target tissues. These tissues do not appear to have an abnormality in Δ^4-5α-reductase activity.

While earlier studies suggested that genital skin or fibroblasts derived from it[162] would be preferrable to nongenital skin for the study of Δ^4-5α-reductase activity in people suspected of having an enzyme deficiency, a recent study on two prepubertal patients with male psuedohermaphroditism supported use of nongenital skin.[163] These two subjects had nongenital skin fibroblast Δ^4-5α-reductase activity levels which were approximately 3% of normal. The low enzyme activity was supported by abnormally high plasma testosterone/5α-dihydrotestosterone levels following gonadotropin stimulation, high urine 5β/5α-testosterone metabolite ratio, deficient conversion of infused ^{3}H-testosterone to 5α-dihydrotestosterone, and lower than normal conversions by genital skin.

The role of Δ^4-5α-reduction of testosterone in certain tissues in the female is unclear. Androgens have been reported to have stimulatory effects on uterine growth in immature rats.[164] Receptors for androgens have been described for uterine tissues for both humans[165] and rats.[166] That Δ^4-5α-reduction might be involved in testosterone effects in the uterus is supported by the finding of Δ^4-5α-reductase activity in tissue from both these species.[167] Some human mammary cancers have 5α-dihydrotestosterone receptors.[165] There have been a number of reports on the conversion of C_{19}-steroids to this active androgen in mammary cancer tissue.[168,169] These two female hormone target tissues are obviously under the control of a number of hormones and the precise role or function of the androgens and hence Δ^4-5α-reductase activity remains to be elucidated.

The Δ^4-5α-reduction of testosterone leads to the formation of 5α-dihydrotestosterone which is the important intracellular androgen in a number of target tissues. Some of the evidence for such has been summarized in particular for the prostate. It has not been detailed for tissues such as the seminal vesicles because it is similar to that for the described tissues. Δ^4-5α-Reduction is probably not involved in some of the behavioral effects of testosterone nor in the effects of testosterone in most types of muscle and in the kidney.

5α-Dihydrotestosterone can be further reduced to the androstanediols (3α and 3β) which may exert actions through conversion back to the same steroid or by different mechanisms. This aspect has not been covered as the 3-hydroxysteroid dehydrogenase is a separate topic.

VI. SUMMARY

Steroid Δ^4-reductases are rate limiting enzymes in the metabolism of corticoids, progesterone, and C_{19}-steroids to products which are completely reduced in ring A and conjugated for excretion. The liver is a major site for this Δ^4-reduction yielding both 5α- and 5β-metabolites.

Δ^4-5α-Reductases are found in certain androgen target tissues and are required for the formation of 5α-dihydrotestosterone which appears to be the active intracellular androgen in tissues such as the prostate. The evidence for the involvement of Δ^4-5α-reduction in the expression of progesterone activity is far less convincing. There is no evidence for its involvement in corticoid actions.

Abnormalities in Δ^4-reductase levels can lead to clinical disorders and several examples have been described.

REFERENCES

1. **Bjorkhem, I.,** Mechanism and stereochemistry of the enzymatic conversion of a Δ^4-3-oxosteroid into a 3-oxo-5α-steroid, *Eur. J. Biochem.,* 8, 345, 1969.
2. **Bjorkhem, I. and Danielsson, H.,** Stereochemistry of hydrogen transfer from pyridine nucleotides catalyzed by Δ^4-3-oxo-steroid 5β-reductase and 3α-hydroxysteroid dehydrogenase from rat liver, *Eur. J. Biochem.,* 12, 80, 1970.
3. **Abul-Hajj, Y. J.,** Stereospecificity of hydrogen transfer from NADPH by steroid Δ^4-5α- and Δ^4-5β-reductase, *Steroids,* 20, 215, 1972.
4. **Bird, C. E. and Clark, A. F.,** The adrenals, in *Systematic Endocrinology,* Ezrin, C., Godden, J. O., Volpe, R., and Wilson, R., Eds., Harper and Row, Hagerstown, N.Y., 1972, 149.
5. **Wortmann, W., Touchstone, J. C., Knapstein, P., Dick, G., and Mappes, G.,** Metabolism of 1,2-^{3}H-cortisol perfused through human liver *in vivo,* J. Clin. Endocrinol. Metab., 33, 597, 1971.
6. **Ganong, W. F.,** The adrenal medulla and adrenal cortex, in *Review of Medical Physiology,* 6th ed., Lange Medical Publications, Los Altos, Calif., 1973, 266.
7. **Fukushima, D. K., Bradlow, H. L., Hellmann, L., Zumoff, B., and Gallagher, T. F.,** Metabolic transformation of hydrocortisone-4-C^{14} in normal men, *J. Biol. Chem.,* 235, 2246, 1960.
8. **Bradlow, H. L., Zumoff, B., Monder, C., Lee, H. J., and Hellman, L.,** Isolation and identification of four new carboxylic acid metabolites of cortisol in man, *J. Clin. Endocrinol. Metab.,* 37, 1973.
9. **Rappaport, R. and Migeon, C. J.,** Physiologic disposition of 4-C^{14}-tetrahydrocortisol in man, *J. Clin. Endocrinol. Metab.,* 22, 1065, 1962.
10. **Fukushima, D. K., Hellman, L., and Gallagher, T. F.,** Reproducibility of a double isotope derivative analysis for cortisol metabolites, *Steroids,* 14, 1, 1969.
11. **Gold, N. I. and Crigler, J. F., Jr.,** Suggested *in vivo* irreversible metabolism of labeled cortisol during distribution: effects on kinetic measurements and role of modes and sites of administration, *J. Clin. Endocrinol. Metab.,* 34, 1025, 1972.
12. **Libert, R. and DeHertogh, R.,** Critical study on cortisol production rate, *Steroids,* 16, 539, 1970.

13. **McGuire, J. S., Jr. and Tomkins, G. M.**, The heterogeneity of Δ^4-3-ketosteroid reductases (5α), *J. Biol. Chem.*, 235, 1634, 1960.
14. **Tomkins, G. M., Bakemeier, R. F., and Weinberg, A. N.**, Biochemistry and biology of the steroid reductases, in *Proc. Int. Congress of Biochemistry*, Vol. 7, Popjak, G., Ed., MacMillan, New York, 1963, 405.
15. **Kitay, J. I.**, Effects of estradiol on pituitary-adrenal function in male and female rats, *Endocrinology*, 72, 947, 1963.
16. **Colby, H. D., Gaskin, J. H., and Kitay, J. I.**, Requirement of the pituitary gland for gonadal hormone effects on hepatic corticosteroid metabolism in rats and hamsters, *Endocrinology*, 92, 769, 1973.
17. **Sandberg, A. A. and Slaunwhite, W. R., Jr.**, Transcortin: a corticosteroid-binding protein of plasma. V. *In vitro* inhibition of cortisol metabolism, *J. Clin. Invest.*, 42, 51, 1963.
18. **Carter, J. R., Hagen, A. A., Biggs, J. T., and Troop, R. C.**, Inhibition of adrenal steroid metabolism by heparin *in vivo*, *Metabolismo*, 17, 352, 1962.
19. **Kornel, L.**, Corticosteroids in human blood. V. Extra-adrenal effect of ACTH upon metabolism of cortisol, *Steroidologia*, 1, 225, 1970.
20. **Yates, F. E., Urquhart, J., and Herbst, A. L.**, Effects of thyroid hormones on ring A reduction of cortisone by liver, *Am. J. Physiol.*, 195, 373, 1959.
21. **Hellman, L., Bradlow, H. L., Zumoff, B., and Gallagher, T. F.**, The influence of thyroid hormone on hydrocortisone production and metabolism, *J. Clin. Endocrinol. Metab.*, 21, 1231, 1961.
22. **New, M. I., Levine, L. S., Biglieri, E. G., Pareira, J., and Ulick, S.**, Evidence for an unidentified steroid in a child with apparent mineralocorticoid hypertension, *J. Clin. Endocrinol. Metab.*, 44, 924, 1977.
23. **Ulick, S., Ramirez, L. C., and New, M. I.**, An abnormality in steroid reductive metabolism in a hypertensive syndrome, *J. Clin. Endocrinol. Metab.*, 44, 799, 1977.
24. **Marver, D. and Edelman, J. S.**, Dihydrocortisol: a potential mineralocorticoid, *J. Steroid Biochem.*, 9, 1, 1978.
25. **Frantz, A. G., Katz, F. H., and Jailer, J. W.**, 6-Beta-hydroxycortisol: high levels in human urine in pregnancy and toxemia, *Proc. Soc. Exp. Biol. Med.*, 105, 41, 1960.
26. **Ulstrom, R. A., Colle, E., Burley, J., and Gunville, R.**, Adrenocortical steroid metabolism in newborn infants. II. Urinary excretion of 6β-hydrocortisol and other polar metabolites, *J. Clin. Endocrinol.*, 20, 1080, 1960.
27. **Layne, D. S., Meyer, C. J., Vaishivanar, P. S., and Pincus, G.**, The secretion and metabolism of cortisol and aldosterone in normal and in steroid-treated women, *J. Clin. Endocrinol.*, 22, 107, 1962.
28. **Bledsoe, T., Island, D. P., Ney, R. L., and Liddle, G. W.**, An effect of o,p′-DDD on the extra-adrenal metabolism of cortisol in man, *J. Clin. Endocrinol. Metab.*, 24, 1303, 1964.
29. **Werk, E. E., MacGee, J., and Sholiton, L. J.**, Effect of diphenylhydantoin on cortisol metabolism in man, *J. Clin. Invest.*, 43, 1824, 1964.
30. **Sowell, J. G., Hagen, A. A., and Troop, R. C.**, Metabolism of cortisone-4-^{14}C by rat lung tissue, *Steroids*, 18, 289, 1971.
31. **Murota, S. I., Endo, H., and Tamaoki, B. I.**, Identification of metabolites of cortisol in cultured bone and their effects upon bone formation, *Biochim. Biophys. Acta*, 136, 379, 1967.
32. **Elattar, T. M. A., Murray, W. R., and Anderson, C. E.**, *In vivo* catabolism of [4-^{14}C] cortisol in synovial fluid of the rheumatoid and osteoarthritic knee, *Metabolism*, 11, 178, 1968.
33. **Reach, G., Nakane, H., Nakane, Y., Auzan, C., and Corvol, P.** Cortisol metabolism and excretion in the isolated perfused rat kidney, *Steroids*, 30, 621, 1977.
34. **Colby, H. D., Malendowicz, L. K., Caffrey, J. L., and Kitay, J. I.**, Role of adrenal 5α-reductase activity in determining the responsiveness of hypophysectomized rats to ACTH, *Endocrinology*, 96, 1153, 1975.
35. **Witorsch, R. J. and Kitay, J. I.**, Pituitary hormones affecting adrenal 5α-reductase activity: ACTH, growth hormone and prolactin, *Endocrinology*, 91, 764, 1972.
36. **Colby, H. D. and Kitay, J. I.**, Effects of gonadal hormones on adrenocortical secretion of 5α-reduced metabolites of corticosterone in the rat, *Endocrinology*, 91, 1523, 1972.
37. **Ogle, T. F.and Kitay, J. I.**, Effects of melatonin and an aqueous pineal extract on adrenal secretion of reduced steroid metabolites in female rats, *Neuroendocrinology*, 23, 113, 1977.
38. **Ogle, T. F. and Kitay, J. I.**, *In vitro* effects of melatonin and serotonin on adrenal steroidogenesis, *Proc. Soc. Exp. Biol. Med.*, 157, 103, 1978.
39. **Kitay, J. I., Coyne, M. D., and Swygert, N. H.**, Adrenal 5α-reductase activity in the rat: effects of gonadectomy: some cofactor and substrate requirements, *Proc. Soc. Exp. Biol. Med.*, 137, 229, 1971.
40. **Flood, C., Pincus, G., Tait, J. F., Tait, S. A. S., and Willoughby, S.**, A comparison of the metabolism of radioactive 17-iso-aldosterone and aldosterone administered intravenously and orally to normal human subjects, *J. Clin. Invest.*, 46, 717, 1967.

41. **Bledsoe, T., Liddle, G. W., Riondel, A., Island, D. P., Bloomfield, D., and Sinclair-Smith, B.,** Comparative fates of intravenously and orally administered aldosterone: evidence for extrahepatic formation of acid-hydrolyzable conjugate in man, *J. Clin. Invest.,* 45, 264, 1966.
42. **Tait, J. F. and Little, B.,** The metabolism of orally and intravenously administered labeled aldosterone in pregnant subjects, *J. Clin. Invest.,* 47, 2423, 1968.
43. **Coppage, W. S., Jr., Island, D. P., Cooner, A. E., and Liddle, G. W.,** The metabolism of aldosterone in normal subjects and in patients with hepatic cirrhosis, *J. Clin. Invest.,* 41, 1672, 1962.
44. **Ulick, S., Kusch, K., and August, J. T.,** Correction of the structure of a urinary metabolite of aldosterone, *J. Am. Chem. Soc.,* 83, 4482, 1961.
45. **Kelly, W. G., Bandi, L., Shoolery, J. N., and Lieberman, S.,** Isolation and characterization of aldosterone metabolites from human urine; two metabolites bearing a bicyclic acetal structure, *Biochemistry,* 1, 172, 1962.
46. **Kelly, W. G., Bandi, L., and Lieberman, S.,** Isolation and characterization of human urinary metabolites of aldosterone. III. Three isomeric tetrahydro metabolites, *Biochemistry,* 1, 792, 1962.
47. **Genest, J., Nowaczynski, W., Boucher, R., and Kuchel, O.,** Role of the adrenal cortex and sodium in the pathogenesis of human hypertension, *Canadian Med. Assoc. J.,* 118, 538, 1978.
48. **Melby, J. C., Dale, S. L., Grekin, R. J., Gaunt, R., and Wilson, T. E.,** 18-Hydroxy-11-deoxycorticosterone (18-OH-DOC) secretion in experimental and human hypertension, *Recent Prog. Horm. Res.,* 28, 287, 1972.
49. **Florini, J. R., Smith, L. L., and Buyske, D. A.,** Metabolic fate of a synthetic corticosteroid (triamcinolone) in the dog, *J. Biol. Chem.,* 236, 1038, 1961.
50. **Florini, J. R. and Buyske, D. A.,** A comparison of the rate of metabolism and biological activity of 16αOH hydrocortisone derivatives, *Arch. Biochem. Biophys.,* 79, 8, 1959.
51. **Venning, E. H. and Browne, J. S. L.,** Isolation of a water-soluble pregnanediol complex from human pregnancy urine, *Proc. Soc. Exp. Biol. Med.,* 34, 792, 1936.
52. **Loraine, J. A.,** Progesterone and its metabolites, in *Clinical Application of Hormone Assay,* E. & S. Livingstone, Edinburgh, 1958, 205.
53. **Judge, S. M., Quade, J. P., Avata, W. S. M., and Chatterton, R. T., Jr.,** Time-course relationships between serum LH, serum progesterone and urinary pregnanediol concentrations in normal women, *Steroids,* 31, 175, 1978.
54. **Romanoff, L. P., Morris, C. W., Welch, P., Grace, M. P., and Pincus, G.,** Metabolism of progesterone-4-^{14}C in young and elderly men, *J. Clin. Endocrinol. Metab.,* 23, 286, 1963.
55. **Mainwaring, W. I. P.,** *The Mechanism of Action of Androgens,* Springer-Verlag, New York, 1977.
56. **Armstrong, D. T. and King, E. R.,** Uterine progesterone metabolism and progestational response: effects of estrogens and prolactin, *Endocrinology,* 89, 191, 1971.
57. **Saffran, J., Loeser, B. K., Haas, B. M., and Stavely, H. E.,** Metabolism of progesterone in subcellular fractions of rat uterus, *Steroids,* 24, 839, 1974.
58. **Verma, V. and Laumas, K. R.,** Effect of estradiol-17β and progesterone on the metabolism of [1,2-^{3}H] progesterone by the rabbit endometrium and myometrium, *Biochem. J.,* 156, 55, 1976.
59. **Rahman, S. S., Billiar, R. B., and Little, B.,** *In vivo* uterine metabolism of progesterone and 5α-pregnane-3,20-dione during the synthesis and release of uteroglobin in ovariectomized, steroid-treated rabbits, *Endocrinology,* 102, 1901, 1978.
60. **Rahman, S. S., Billiar, R. B., and Little, B.,** Induction of uteroglobin in rabbits by progestegens, estradiol 17β and ACTH, *Biol. Reprod.,* 12, 305, 1975.
61. **Saffran, J., Loeser, B. K., Bohnett, S. A., Gray, M. A., and Faber, L. E.,** The binding of 5α-pregnane-3,20-dione by cytosol and nuclear preparations of guinea pig uterus, *Endocrinology,* 192, 1088, 1978.
62. **Milgrom, E., Thi, L., Atger, M., and Baulieu, E.-E.,** Mechanism regulating the concentration and conformation of progesterone receptor(s) in the uterus, *J. Biol. Chem.,* 248, 6366, 1973.
63. **Milgrom, E., Atger, M., and Baulieu, E.-E.,** Progesterone in uterus and plasma. IV. Progesterone receptor(s) in guinea pig uterus cytosol, *Steroids,* 16, 741, 1970.
64. **Junkerman, H., Runnebaum, B., and Lisboa, B. P.,** New progesterone metabolites in human myometrium, *Steroids,* 30, 1, 1977.
65. **Karavolas, H. J., Hodges, D., and O'Brien, D. O.,** Uptake of [^{3}H]-progesterone and [^{3}H]-5α-dihydroprogesterone by rat tissues *in vivo* and analysis of accumulated radioactivity: accumulation of 5α-dihydroprogesterone by pituitary and hypothalamic tissues, *Endocrinology,* 98, 164, 1976.
66. **Whalen, R. E. and Gorzalka, B. B.,** The effects of progesterone and its metabolites on the induction of sexual receptivity in rats, *Horm. and Behav.,* 3, 221, 1972.
67. **Lee, D. K. H., Bird, C. E., and Clark, A. F.,** *In vivo* metabolism of ^{3}H-testosterone in adult male rats: effects of estrogen administration, *Steroids,* 26, 137, 1975.
68. **Van Doorn, E. J., Burns, B., Wood, D., Bird, C. E., and Clark, A. F.,** *In vivo* metabolism of ^{3}H-dihydrotestosterone and ^{3}H-androstanediol in adult male rats, *J. Steroid Biochem.,* 6, 1549, 1975.

69. **Karavolas, H. J. and Herf, S. M.**, Conversion of progesterone by rat medial basal hypothalamic tissue to 5α-pregnane-3,20-dione, *Endocrinology,* 89, 940, 1971.
70. **Hanukoglu, I., Karavolas, H. J., and Gray, R. W.**, Progesterone metabolism in the pineal, brain stem, thalamus and corpus collosum of the female rat, *Brain Res.,* 125, 313, 1977.
71. **Little, B., Billiar, R. B., Rahman, S. S., Johnson, W. A., Takaoka, Y., and White, R. J.**, *In vivo* aspects of progesterone distribution and metabolism, *Am. J. Obstet. Gynecol.,* 123, 527, 1975.
72. **Mauvais-Jarvis, P., Baudot, N., and Bercovici, J. P.**, *In vivo* studies on progesterone metabolism by human skin, *J. Clin. Endocrinol. Metab.,* 29, 1580, 1969.
73. **Gomez, E. C., Frost, P., and Hsia, S. L.**, Transformations of progesterone by subcellular fractions of human skin, *J. Invest. Dermatol.,* 64, 240, 1975.
74. **Mauvais-Jarvis, P., Baudot, N., and Bercovici, J. P.**, *In vivo* studies on progesterone metabolism by human skin, *J. Clin. Endocrinol. Metabl.,* 29, 1580, 1969.
75. **Mercier-Bodard, C., Marchut, M., Perrot, M., Picard, M.-T., Balieu, E.-E., Robel, P.**, Influence of purified plasma proteins on testosterone uptake and metabolism by normal and hyperplastic human prostate in "Constant-flow organ culture," *J. Clin. Endocrinol. Metab.,* 43, 374, 1976.
76. **Milewich, L., Gomez-Sanchez, C., Crowley, G., Porter, J. C., Madden, J. D., and MacDonald, P. C.**, Progesterone and 5α-pregnane-3,20-dione in peripheral blood of normal young women. Daily measurements throughout the menstrual cycle, *J. Clin. Endocrinol. Metab.,* 45, 617, 1977.
77. **Pollow, K., Liibbert, H., Boquoi, E., Pollow, B.**, Progesterone metabolism in normal human endometrium during the menstrual cycle and in endometrial carcinoma, *J. Clin. Endocrinol. Metab.,* 41, 729, 19.
78. **Schwarz, B. E., Milewich, L., Grant, N. F., Porter, J. C., Johnston, J. M., and MacDonald, P. C.**, Progesterone binding and metabolism in human fetal membranes, *Ann. N.Y. Acad. Sci.,* 286, 304, 1977.
79. **Sridharan, B. N., Meyer, R. K., and Karavolas, H. J.**, Effect of 5α-dihydroprogesterone, pregn-5-ene-3,20-dione, pregnenolone and related progestins on ovulation in PMSG-treated immature rats, *J. Reprod. Fertil.,* 36, 83, 1974.
80. **Wielgosz, G. J. and Armstrong, D. T.**, Effects of prolactin and progesterone on preputial gland growth and progesterone metabolism in hypophysectomized, ovariectomized immature rats, *J. Steroid Biochem.,* 8, 1051, 1977.
81. **Chatterton, R. T., Jr., Chatterton, A. J., and Hellman, L.**, Metabolism of progesterone by the rabbit mammary gland, *Endocrinology,* 85, 16, 1969.
82. **Podratz, K. C., Munns, T. W., and Katzman, P. A.**, Metabolism of progesterone in mouse vaginal tissue, *Steroids,* 24, 775, 1974.
83. **O'Malley, B. W., Sherman, M. R., Toft, D. O., Spelsberg, T. C., Shrader, W. T., and Steggles, A. W.**, A specific oviduct target-tissue receptor for progesterone, in *Advances in the Biosciences,* Vol. 7, Raspe, G., Ed., Pergamon-Vieweg, Oxford, 1970, 213.
84. **Morgan, M. D. and Wilson, J. D.**, Intranuclear metabolism of progesterone-1,2-^{3}H in the hen oviduct, *J. Biol. Chem.,* 245, 3781, 1970.
85. **Dorfman, R. I., and Shipley, R. A.**, Relative activities of androgens, in *Androgens: Biochemistry, Physiology and Chemical Significance,* John Wiley & Sons, New York, 1956, 116.
86. **Vande Wiele, R. L., MacDonald, P. C., Gurpide, E., and Lieberman, S.**, Studies on the secretion and interconversion of the androgens, *Recent Prog. Horm. Res.,* 19, 275, 1963.
87. **Horton, R., Shinsako, J., and Forsham, P. H.**, Testosterone production and metabolic clearance rates with volumes of distribution in normal men and women, *Acta Endocrinol. (Copenhagen),* 48, 446, 1965.
88. **de Nicola, A. F., Dorfman, R. I., and Forchielli, E.**, Urinary excretion of epitestosterone and testosterone in normal individuals and hirsute and virilized females, *Steroids,* 7, 351, 1966.
89. **Clark, A. F., Carson, G. D., DeLory, B., Clemow, M. E., and Bird, C. E.**, Effects of estrogen administration on testosterone metabolism in normal men, *Clin. Endocrinol. (Oxford),* 2, 361, 1973.
90. **Gordon, G. G., Altman, K., Southren, A. L., and Olivo, J.**, Human hepatic testosterone A ring reductase activity: effect of medroxyprogesterone acetate, *J. Clin. Endocrinol. Metab.,* 32, 457, 1971.
91. **Gordon, G. G., Southren, A. L., Tochimoto, S., Olivo, J., Altman, K., Rand, J., and Lemberger, L.**, Effect of medroxyprogesterone acetate (Provera®) on the metabolism and biological activity of testosterone, *J. Clin. Endocrinol. Metab.,* 30, 449, 1970.
92. **Bird, C. E. and Clark, A. F.**, Kinetics of ^{3}H-5α-dihydrotestosterone metabolism in normal men and women and men with prostatic carcinoma: effects of estrogen administration in men, *J. Clin. Endocrinol. Metab.,* 34, 467, 1972.
93. **Clark, A. F., Calandra, R. S., and Bird, C. E.**, Binding of testosterone and 5α-dihydrotestosterone to plasma proteins in humans, *Clin. Biochem.,* 4, 89, 1973.
94. **Murphy, B. E. P.**, Binding of testosterone and estradiol in plasma, *Can. J. Biochem.,* 46, 299, 1968.
95. **Zumoff, B., Bradlow, H. L., Finkelstein, J., Boyar, R. M., and Hellman, L.**, The influence of age and sex on the metabolism of testosterone, *J. Clin. Endocrionol. Metab.,* 42, 703, 1976.

96. **Bird, C. E. and Clark, A. F.**, unpublished observations.

97. **Nduaguba, J. C. and Clark, A. F.**, Steroid Δ^4-reductases of pig liver: partial purification of testosterone 5β-reductase, *Can. J. Biochem.*, 49, 385, 1971.

98. **Patterson, D. C., Clark, A. F., and Bird, C. E.**, Testosterone Δ^4-reductase activity in rat liver: hormonal control *in vivo*, J. Endocrinol., 63, 181, 1974.

99. **Bird, C. E., Green, R. N., Calandra, R. S., Connolly, J. G., and Clark, A. F.**, Kinetics of ^{3}H-testosterone metabolism in patients with carcinoma of the prostate: effects of oestrogen administration, *Acta Endocrinol. (Copenhagen)*, 67, 733, 1971.

100. **Kato, R., Onoda, K., and Omori, Y.**, Mechanism of thyroxine-induced increase in steroid Δ^4-reductase activity in male rats, *Endocrinol. Jpn.*, 17, 215, 1970.

101. **Hellman, L., Bradlow, H. L., Zumoff, B., Fukushima, D. K., and Gallagher, T. F.**, Thyroid-androgen interrelations and the hypercholesteremic effect of androsterone, *J. Clin. Endocrinol. Metab.*, 19, 936, 1959.

102. **Stephens, J. K., Fischer, P. W. F., and Marks, G. S.**, Porphyrin induction: equivalent effects of 5αH and 5βH steroids in chick embryo liver cells, *Science*, 197, 659, 1977.

103. **Kappas, A., Bradlow, H. L., Gillette, P. N., and Gallagher, T. F.**, Abnormal steroid metabolism in the genetic liver disease, acute intermittent porphyria, *Ann. N.Y. Acad. Sci.*, 179, 611, 1971.

104. **Kappas, A., Bradlow, H. L., Bukers, D. R., and Alvares, A. P.**, Induction of a deficiency of steroid Δ^4-5α-reductase activity in liver by a porphyrinogen drug, *J. Clin. Invest.*, 59, 159, 1977.

105. **Gustafsson, J. A., Emeroth, P., Pousette, A., Skett, P., Sonnenschein, C., Stenberg, A., and Ahlin, A.**, Programming and differentiation of rat liver enzymes, *J. Steroid Biochem.*, 8, 429, 1977.

106. **Gustafsson, J. A., Pousette, A., and Svensson, E.**, Sex-specific occurrence of androgen receptors in rat brain, *J. Biol. Chem.*, 251, 4047, 1976.

107. **Pearlman, W. H. and Pearlman, M. R. J.**, The metabolism *in vivo* of Δ^4-androstene-3,17-dione-7-H^3, its localization in the ventral prostate and other tissues of the rat, *J. Biol. Chem.*, 236, 1321, 1961.

108. **Bruchovsky, N., and Wilson, J. D.**, The conversion of testosterone to 5α-androstan-17β-ol-3-one by rat prostate *in vivo* and *in vitro*, J. Biol. Chem., 243, 2012, 1968.

109. **Anderson, K. M. and Liao, S.**, Selective retention of dihydrotestosterone by prostatic nuclei, *Nature (London)*, 219, 277, 1968.

110. **Baulieu, E.-E., Lasnitzki, I., and Robel, P.**, Testosterone, prostate gland and hormone action, *Biochem. Biophys. Res. Commun.*, 32, 575, 1968.

111. **Bruchovsky, N.**, Comparison of the metabolites formed in the rat prostate following the *in vivo* administration of seven natural androgens, *Endocrinology*, 89, 1212, 1971.

112. **Wilson, J. D. and Gloyna, R. E.**, The intranuclear metabolism of testosterone in the accessory organs of reproduction, *Recent Prog. Horm. Res.*, 26, 309, 1970.

113. **Frederiksen, D. W. and Wilson, J. D.**, Partial characterization of the nuclear reduced nicotinamide adenine dinucleotide phosphate: Δ^4-3-ketosteroid 5α-oxidoreductase of rat prostate, *J. Biol. Chem.*, 246, 2584, 1971.

114. **Moore, R. J. and Wilson, J. D.**, Localization of the reduced nicotinamide adenine dinucleotide phosphate: Δ^4-3-ketosteroid 5α-oxidoreductase in the nuclear membrane of the rat ventral prostate, *J. Biol. Chem.*, 247, 958, 1972.

115. **Prout, C. R., Sierp, M., and Whitmore, W. F.**, Radioactive zinc in the prostate, *JAMA*, 169, 1703, 1959.

116. **Lee, D. K. H., Bird, C. E., and Clark, A. F.**, Studies on the *in vivo* hormonal control of rat prostatic testosterone 4-ene-5α-reductase activity, *J. Steroid Biochem.*, 5, 609, 1974.

117. **Gloyna, R. E., Siiteri, P. K., and Wilson, J.**, Dihydrotestosterone in prostatic hypertrophy. II. The formation and content of dihydrotestosterone in the hypertrophic canine prostate and the effect of dihydrotestosterone on prostatic growth in the dog, *J. Clin. Invest.*, 49, 1746, 1970.

118. **Walsh, P. C. and Wilson, J. D.**, The induction of prostatic hypertrophy in the dog with androstanediol, *J. Clin. Invest.*, 57, 1093, 1976.

119. **Ishimaru, T., Pages, L., and Horton, R.**, Altered metabolism of androgens in elderly men with benign prostatic hyperplasia, *J. Clin. Endocrinol. Metab.*, 45, 697, 1977.

120. **Haltmeyer, G. C. and Eik-Nes, K. B.**, Production and secretion of 5α-dihydrotestosterone by the canine prostate, *Acta Endocrinol. (Copenhagen)*, 69, 394, 1972.

121. **Vermeulen, A. and DeSy, W.**, Androgens in patients with benign prostatic hyperplasia, *J. Clin. Endocrinol. Metab.*, 43, 1250, 1976.

122. **Cowan, R. A., Cowan, S. K., Grant, J. K., and Elder, H. Y.**, Biochemical investigations of separated epithelium and stroma from benign hyperplastic prostatic tissue, *J. Endocrinol.*, 74, 111, 1977.

123. **Habib, F. K., Lee, J. R., Stitch, S. R., and Smith, P. H.**, Androgen levels in the plasma and prostatic tissues of patients with benign hypertrophy and carcinoma of the prostate, *J. Endocrinol.*, 71, 99, 1976.

124. **Giorgi, E. P., Stewart, J. C., Grant, J. K., and Shirley, I. M.,** Androgen dynamics *in vitro* in the human prostate gland. Effects of estradiol-17β, *Biochem. J.*, 126, 107, 1972.
125. **Geller, J., Albert, J., Loza, D., Geller, S., Stoeltzing, W., and de la Vega, D.,** DHT concentrations in human prostate cancer tissue, *J. Clin. Endocrinol. Metab.*, 46, 440, 1978.
126. **Farnsworth, W. E. and Brown, J. R.,** Androgens of the human prostate, *Endocr. Res. Commun.*, 3, 105, 1976.
127. **Baker, H. W. G., Burger, H. G., de Kretser, D. M., Hudson, B., and Straffon, W. G.,** Effects of synthetic oral estrogens in normal men and patients with prostatic carcinoma: lack of gonadotrophin suppression by chlorotrianisene, *Clin. Endocrinol; (Oxford)*, 2, 297, 1973.
128. **Yanaihara, T., and Troen, P.,** Studies of the human testes. III. Effect of estrogen on testosterone formation in human testes *in vitro*, J. Clin. Endocrinol. Metab., 34, 968, 1972.
129. **Lee, D. K. H., Bird, C. E., and Clark, A. F.,** *In vitro* effects of estrogens on rat prostatic 5α-reductase of testosterone, *Steroids*, 22, 677, 1973.
130. **Prout, G. R., Jr., Kliman, B., Daly, J. J., MacLaughlin, R. A., and Griffin, P. P.,** *In vitro* uptake of ^{3}H-testosterone and its conversion to dihydrotestosterone by prostatic carcinoma and other tissues, *J. Urol.*, 116, 603, 1976.
131. **Kadohama, N., Kirdini, R. Y., Murphy, G. P., and Sandberg, A. A.,** 5α-Reductase as a target enzyme for antiprostatic drugs in organ culture, *Oncology*, 34, 123, 1977.
132. **Shain, S. A., McCullough, B., Nitchuk, M., and Boesel, R. W.,** Prostate carcinogenesis in the AXC rat, *Oncology*, 34, 114, 1977.
133. **Rivarola, M. A. Podesta, E. J., Chemes, H. E., and Agular, D.,** In vitro metabolism of testosterone by whole human testis, isolated seminiferous tubules and interstitial tissue, *J. Clin. Endocrinol. Metab.*, 37, 454, 1973.
134. **Tence, M. and Drosdowsky, M.,** Biosynthesis and metabolism of testosterone by Sertoli cell-enriched seminferous tubules, *Biochem. Biophys. Res. Comm.*, 73, 47, 1976.
135. **Purvis, K., Calandra, R., Haug, E., and Hansson, V.,** 5α-Reduced androgens and testicular function in the immature rat: effects of 5α-androstan-17β-ol-3-one (DHT) propionate and 5α-androstan-3α,17β-diol, *Mol. Cell. Endocrinol.*, 7, 203, 1977.
136. **Harris, M. E., Bartke, A., Weisz, J., and Watson, D.,** Effects of testosterone and dihydrotestosterone on spermatogenesis, rete testis fluid and peripheral androgen levels in hypophysectomized rats, *Fertil. and Steril.*, 28, 1113, 1977.
137. **Ficher, M. and Steinberg, E.,** *In vitro* progesterone metabolism by rat testicular tissue at different stages of development, *Acta Endocrinol. (Copenhagen)*, 68, 285, 1971.
138. **Lacroix, E., Echaute, W., and Leusen, I.,** Influence of age on the formation of 5α-androstanediol and 7α-hydroxytestosterone by incubated rat testes, *Steroids*, 25, 649, 1975.
139. **Chubb, C., Ewing, L., Irby, D., and Desjardins, C.,** Testicular maturation in the rabbit: secretion of testosterone, dihydrotestosterone, 5α-androstane-3α,17β-diol and 5α-androstane-3β,17β-diol by perfused rabbit testes, epididymis and spermatogenesis, *Biol. Reprod.*, 18, 212, 1978.
140. **Monsalve, A. and Blaquer, J. A.,** Partial characterization of epididymal 5α-reductase in the rat, *Steroids*, 30, 41, 1977.
141. **Robaire, B., Ewing, L. L., Zirkin, B. R., and Irby, D. C.,** Steroid Δ^4-5α-reductase and 3α-hydroxysteroid dehydrogenase in the rat epididymis, *Endocrinology*, 101, 1379, 1977.
142. **Pujol, A. and Bayard, F.,** 5α-Reductase and 3α-hydroxysteroid oxidoreductase enzyme activities in epididymis and their control by androgen and rete testes fluid, *Steroids*, 31, 485, 1978.
143. **Naftolin, F., Ryan, K. J., Davies, I. J., Reddy, V. V., Flores, F., Petro, Z., and Kuhn, M.,** The formation of estrogens by central neuroendocrine tissues, *Recent Prog. Horm. Res.*, 31, 295, 1975.
144. **Jenkins, J. S. and Hall, C. J.,** Metabolism of [^{14}C] testosterone by human fetal and adult brain tissue, *J. Endocrinol.*, 74, 425, 1977.
145. **Massa, R. R., Stupnicka, E., Knicwald, Z., and Martini, L.,** The transformation of testosterone into dihydrotestosterone by the brain and anterior pituitary, *J. Steroid Biochem.*, 3, 385, 1972.
146. **Weisz, J. and Gills, C.,** Metabolites of testosterone in the brain of the newborn female rat after an injection of tritiated testosterone, *Neuroendocrinology*, 14, 72, 1974.
147. **Jaffe, R. B.,** Testosterone metabolism in target tissues: hypothalamic and pituitary tissues of the adult rat and human fetus and the immature rat epiphysis, *Steroids*, 14, 483, 1969.
148. **Lieberburg, I. and McEwen, B. S.,** Brain cell nuclear retention of testosterone metabolites, 5α-dihydrotestosterone and estradiol-17β in adult rats, *Endocrinology*, 100, 588, 1977.
149. **Thien, N. C., Duval, J., Samperez, S., and Jovan, P.,** Testosterone 5α-reductase of microsomes from rat anterior hypophysis: properties, increase by castration and hormonal control, *Biochimie*, 56, 899, 1974.
150. **Verjans, H. L. and Eik-Nes, K. B.,** Effects of androstenes, 5α-androstanes, 5β-androstanes, oestrones and oestratrienes on serum gonadotropin levels and ventral prostate weights in gonadectomized adult male rats, *Acta Endocrinol. (Copenhagen)*, 83, 201, 1976.

151. **Eldridge, J. C. and Mahesh, V. B.**, Pituitary-gonadal axis before puberty: evaluation of testicular steroids in the male rat, *Biol. Reprod.*, 11, 385, 1974.
152. **Hamburger-Bar, R. and Rigter, H.**, Peripheral and central androgenic stimulation of sexual behaviour of castrated male rats, *Acta Endocrinol. (Copenhagen)*, 84, 813, 1977.
153. **Soderstein, P. and Hansen, S.**, Effects of castration and testosterone, dihydrotestosterone or estradiol replacement treatment in neonatal rats on mounting behaviour in the adult, *J. Endocrinol.*, 76, 251, 1978.
154. **Steele, R. E., Didato, F., and Steinetz, B. G.**, Relative importance of 5α-reduction for the androgenic and LH-inhibiting activities of Δ^4-3-ketosteroids, *Steroids*, 29, 331, 1977.
155. **Hodgins, M. B. and Hay, J. B.**, Steroid metabolism in human skin: its relation to sebaceous gland growth and acne vulgaris, *Biochem. Soc. Trans.*, 4, 605, 1976.
156. **Kuttenn, F. and Mauvais-Jarvis, P.**, Testosterone 5α-reduction in the skin of normal subjects and of patients with abnormal sex development, *Acta Endocrinol. (Copenhagen)*, 79, 164, 1975.
157. **Siiteri, P. K. and Wilson, J. D.**, Testosterone formation and metabolism during male sexual differentiation in the human embryo, *J. Clin. Endocrinol. Metab.*, 38, 113, 1974.
158. **Imperato-McGinley, J., Guerrero, L., Gauter, T., and Peterson, R. E.**, Steroid 5α-reductase deficiency in man: an inherited form of male pseudohermaphroditism, *Science*, 186, 1213, 1975.
159. **Walsh, P. C., Madden, J. D., Harrod, M. J., Goldstein, J. L., MacDonald, P. C., and Wilson, J. D.**, Familial incomplete male psuedohermaphroditism Type 2, *N. Engl. J. Med.*, 291, 944, 1974.
160. **Kaufman, M., Strausfeld, C., and Pinsky, L.**, Male pseudohermaphroditism to androgens: deficient 5α-dihydrotestosterone binding in cultured skin fibroblasts, *J. Clin. Invest.*, 58, 345, 1976.
161. **Bardin, C. W., Bullock, L. P., Sherins, R. J., Mowszowicz, I., and Blackburn, W. R.**, Androgen metabolism and mechanism of action in male psuedohermaphroditism: a study of testicular feminization, *Recent Prog. Horm. Res.*, 29, 65, 1973.
162. **Pinsky, L., Kaufman, M., Lambert, B., Faucher, G., and Rosenfeld, R.**, Male pseudohermaphroditism: diagnosis in cell culture, *Can. Med. Assoc. J.*, 116, 1274, 1977.
163. **Saenger, P., Goldman, A. S., Levine, L. S., Korthschutz, S., Meucke, E. C., Katsumata, M., Doberne, Y., and New, M. I.**, Prepubertal diagnosis of steroid 5α-reductase deficiency, *J. Clin. Endocrinol. Metab.*, 46, 627, 1978.
164. **Lerner, L. J.**, Hormones antagonists: inhibitors of specific activities of estrogens and androgens, *Recent Prog. Horm. Res.*, 20, 435, 1964.
165. **Poortman, J., Prenen, J. A. C., Schwarz, F., and Thyssen, J. H. H.**, Interaction of Δ^5-androstene-3β,17β-diol with estradiol and dihydrotestosterone receptors in human myometrial and mammary cancer tissue, *J. Clin. Endocrinol. Metab.*, 40, 373, 1975.
166. **Gianopoulous**, G. Binding of testosterone to cytoplasmic components of the immature rat uterus, *Biochem. Biophys. Res. Commun.*, 44, 943, 1971.
167. **Rose, L. I., Reddy, V. V., and Biondi, R.**, Reduction of testosterone to 5α-dihydrotestosterone by human and rat uterine tissues, *J. Clin. Endocrinol. Metab.*, 46, 766, 1978.
168. **Miller, W. R., MacDonald, D., MacFadyen, I., Roberts, M. M., and Forrest, A. P. M.**, Androgen metabolism in gynaecomastic breast tissue, *Clin. Endocrinol. (Oxford)*, 3, 123, 1974.
169. **Miller, W. R.**, Hormonal status and testosterone metabolism of DMBA-inducted rat mammary carcinomas, *Br. J. Cancer*, 34, 296, 1976.

Chapter 2

STEROID CARBOXYLIC ACIDS

H. L. Bradlow and C. Monder

TABLE OF CONTENTS

I. GENERAL INTRODUCTION

Although the importance of the carboxylic acid group is acknowledged in the sterols, it is generally presumed that in the biologically significant steroid hormones the carboxylic acid function plays no role. The metabolism of the major steroid hormones, the androgens, estrogens, gestogens, or corticosteroids, though directed towards increased polarity, reach this goal by modifications involving the introduction of hydroxyl groups *de novo* or by reduction of extant carbonyl groups and by further selective conjugation of the steroid alcohols with sulfuric or glucuronic acids. In no other group of biological substances has the carboxyl group been considered to participate to so negligible an extent. The results of several research groups suggest that a reassessment of the place of the carboxyl group in steroid metabolism is justified at this time. In this review we will examine how the carboxylic acids participate in sterol metabolism and function. The principle thrust of this survey will be to cover the newly discovered acidic metabolites of the corticosteroids. There is little doubt that this new class of steroids will become of major physiological and clinical interest in years to come. Therefore, it is appropriate to present an overview of the field in this early stage of its development. A thorough survey of this complex area should properly include the glycones and lactones, such as the cardiolactones and the toad poisons. In the interests of brevity, these classes of compounds have been excluded and the discussion is limited to those steroid metabolites containing the carboxylic acid function as an intrinsic part of the steroid, not as part of a conjugate.

II. ACID INTERMEDIATES IN CHOLESTEROL BIOSYNTHESIS

A. Demethylation at C-4

Lanosterol, the earliest sterol to emerge from squalene, is molded and reshaped during its metabolism to serve many functions. During its transformations, important end products and transitory intermediates are formed whose properties and physiological roles are strongly influenced by, and sometimes controlled by, the presence in them of hydroxyl, carbonyl, olefin, and carboxylic acid groups.

The biosynthesis of cholesterol starts with acetate and leads through mevalonate to squalene, which condenses to form lanosterol. Further transformations result in the loss of two carbons at C-4 and one carbon at C-14. A C_{30} sterol (I) is thus transformed into the C_{27} sterol, cholesterol (II). The methyl groups at C-4, shown below,

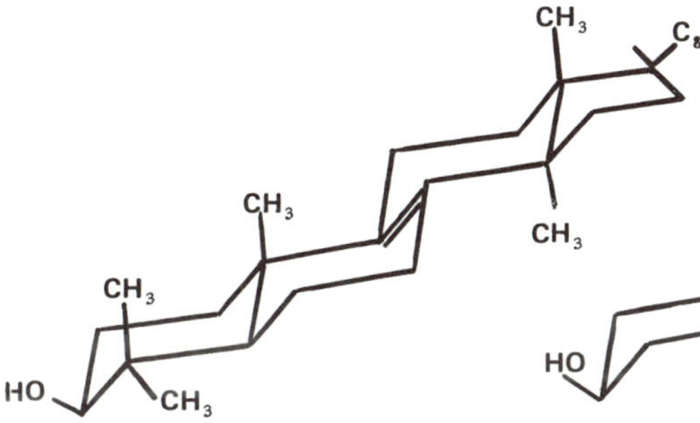

CH3 C8H15 CH3 CH3 CH3 HO CH3 CH3 C8H17 CH3 HO

I 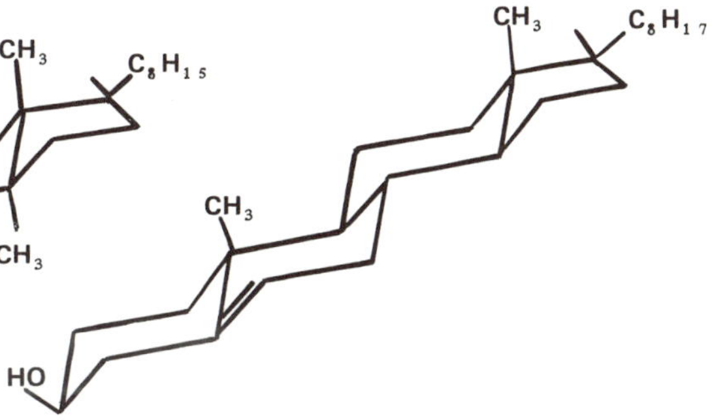 II

are eliminated as carbon dioxide[1] by the action of microsomal enzymes. When microsomes are depleted of NAD^+ and supplemented with oxygen and NADPH, the 4α carboxylic acid intermediate accumulates,[2] as Bloch had predicted earlier.[3,4] The 30-CH_3 and 31-CH_3 are lost sequentially, with the α-methyl group[5] going first. The final stages of oxidation and decarboxylation require NAD^+. The probable sequence is illustrated in Figure 1.

The basic mechanism involves the decarboxylation of a sterol 3-oxo-4α-carboxylic acid.[6-8] Both methyl groups are lost by β-elimination of a carboxyl group resulting from the enzymic oxidation of the α-oriented methyl group. Whether or not a P-450-dependent system is involved is still not established.[9] The conclusion of Mendelsohn et al.[10] that a conventional P-450-dependent component is not present supports the findings of Bechtold et al.,[11] that the methyl steroid oxidase of liver microsomes is an unusual NADH-dependent mixed function oxidase, possibly using an NAD^+-dependent cytochrome b_5 reducing system.

The same oxidase is responsible for the stereospecific demethylation of both α and β methyl groups. After the loss of the first methyl group, the second reorients by a nonenzymic epimerization to yield a 4α methyl-3-ketone. The second cycle of decarboxylation is initiated only after the 3-ketone has been reduced to a β-hydroxy group with a NADPH-dependent 3-ketosteroid reductase.

The Bloch-Gaylor mechanism is supported by considerable indirect evidence: loss of tritium from labeled intermediates,[12,13] identification of intermediates accumulating in depleted systems,[14] the effects of specific cofactors on the metabolism of specific intermediates,[15,16] lack of conversion of proposed alternate intermediates,[17] and stoichiometry of uptake of oxygen and oxidation of NADH.[18]

During the sequence leading to the loss of methyl groups at C-4 and C-14, the double bond at position 8 is retained. Isomerization of 7-ene to 8-ene follows the loss of the methyl groups. The 4-demethylation of cholest-7-ene-3β-ol can occur, but this is not the natural substrate. Competitive inhibition of the decarboxylation by cyclic AMP has been shown. A polar intermediate, 4β-methylzymosterol-4α-carboxylic acid, accumulated.[12] This observation is curious, for cyclic AMP has been shown to stimulate

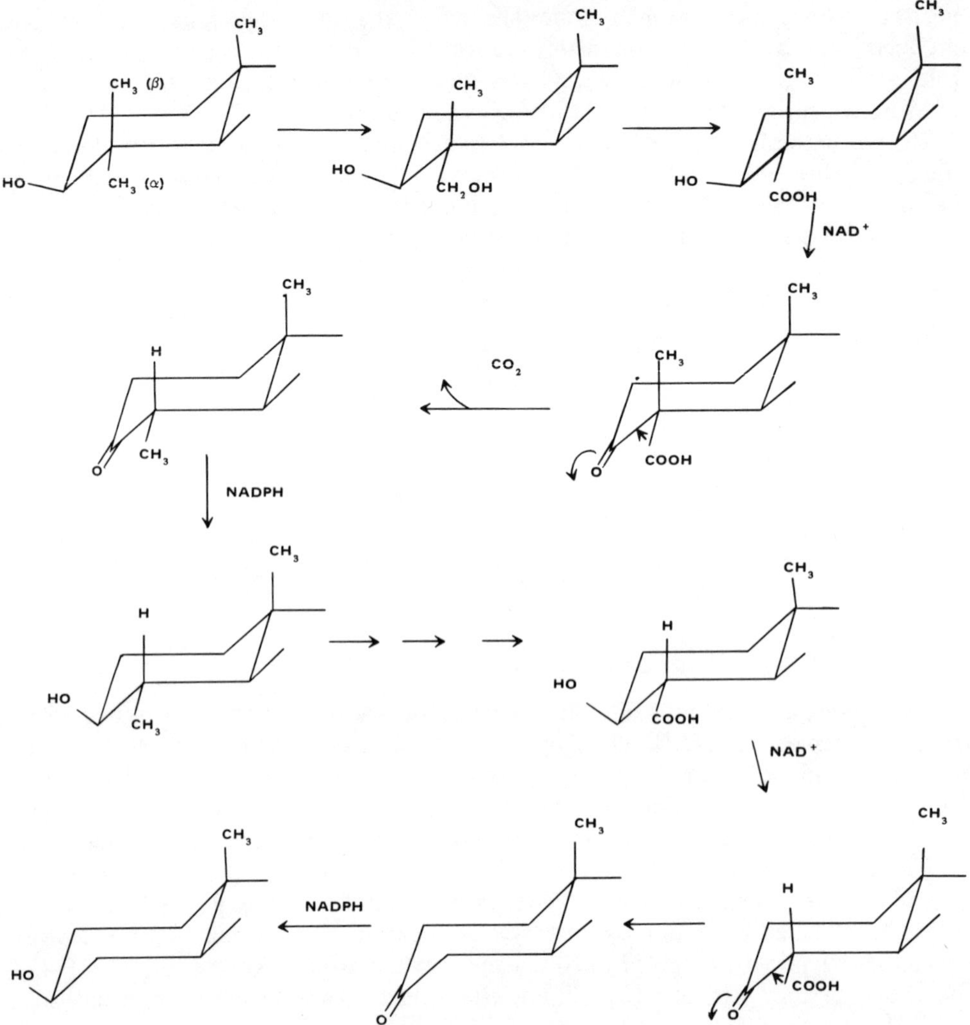

FIGURE 1. Conversion of lanosterol to cholesterol: demethylation at C-4.

cholesterol biosynthesis in vivo. Furthermore, Gaylor and Delwiche[19] have reported that human chorionic gonadotropin, which acts via cyclic AMP as second messenger, stimulated microsomal methyl sterol demethylase in vivo. These contradictory results may reflect the differences in the experimental approaches — the former with tissue homogenates and the latter with intact animals. Demethylases are found wherever cholesterol is synthesized. They have been demonstrated in yeast, skin, testis, and liver.[20] Though most studies have been with liver, all the enzymes show similar characteristics. They require neutral pH, NAD^+, NADPH, and oxygen. The systems are insensitive to carbon monoxide, but are inhibited by cyanide.

B. Demethylation at C-14

Removal of the 32-methyl group at C-14 occurs by a different mechanism than that at C-4. It is included here because it had been considered that C-14 demethylation may involve the C-14 carboxylic acid, lanosteroic acid, as an intermediate.[21-24] The C-32 methyl group is currently believed to be oxidized to aldehyde and lost as formic acid[25,26] in a system containing unwashed microsomes and NAD^+. The overall reaction requires NADPH and oxygen, suggesting the need for a P-450-dependent system.

III. STEROID ACIDS FROM LOWER ORDERS

A number of steroidal acids containing the carboxyl group on the side chain have been isolated from fungi and molds. *Eburicoic acid* has been isolated from *Basidiomycetes.*[27,28] The fungus from which it was first isolated, and from which its name is derived, is the Japanese ''eburico'' or *Fomes officinalis Fris.* It is present as a conjugate.[29] Accompanying it is a dehydro derivative, identified as a 7(8), 10(11) diene.

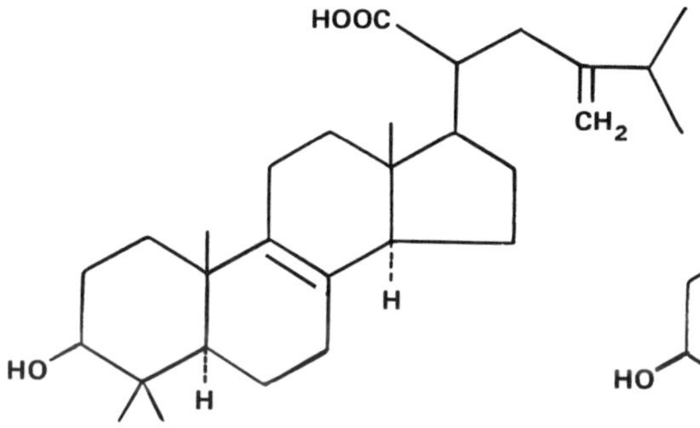

Eburicoic acid

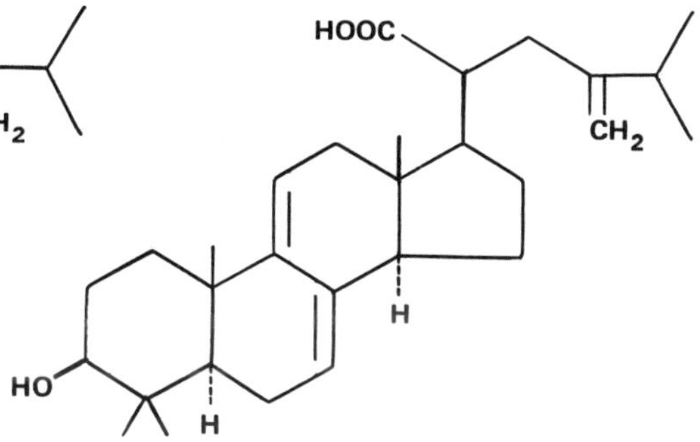

Dehydroeburicoic acid

The acid, found also in the shelf fungus, *Polyporus sulphurens,* is accompanied by its 15α-hydroxy derivative, sulphurenic acid.[30]

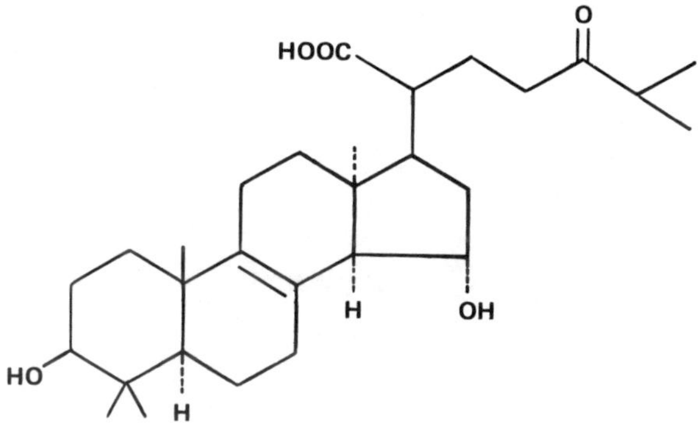

Sulfurenic acid

Eburicoic acid is structurally similar to the antibiotics cephalosphorin P, helvolic acid, and fusidic acid. All are tetracyclic triterpenoids with 21-carboxyl groups.

In addition to these acids, a number of sterols, the polyporenic acids, have been isolated from the genus *Polyporus.* Most studies have been performed by Jones et al.[31-35] Polyporenic acid was isolated from *Polyporus betulinus* by Frerejaque.[36] This group of sterols has antibacterial actions.[37] Polyporenic acids A, B, and C were separated after saponification of fungal extracts by Cross et al.[38] Polyporenic acid A is a 26-carboxy sterol.

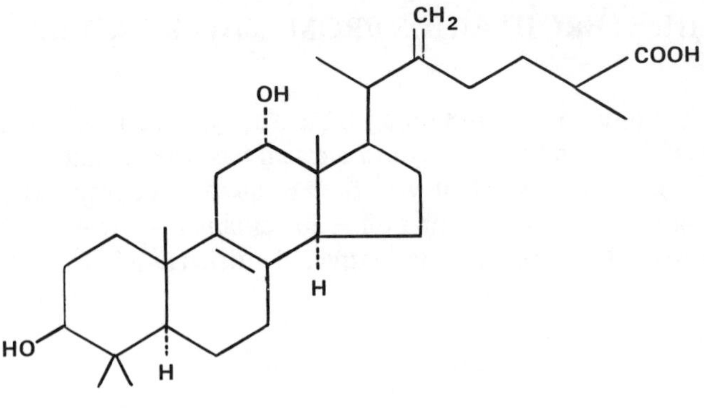

Polyporenic Acid A

Because of the β-methylenic carbon, it is readily decarboxylated when heated, and the methylene group migrates to the 24:25 position.[39]

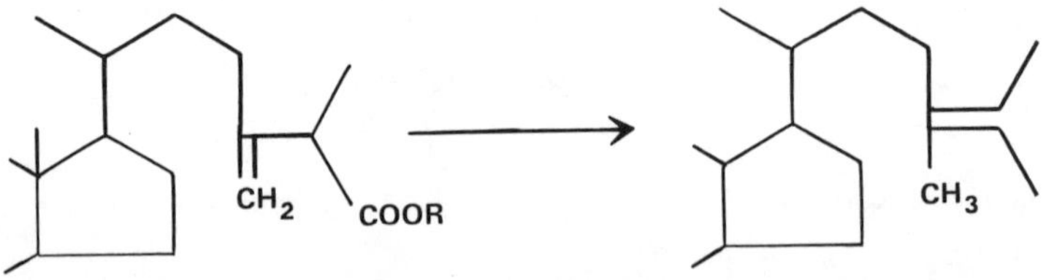

Polyporenic acid B is a mixture of tumulosic acid and dehydrotumulosic acid. Tumulosic acid differs from polyporenic acid only in the position of a ring hydroxy group, which is 16α in the former and 12α in the latter.

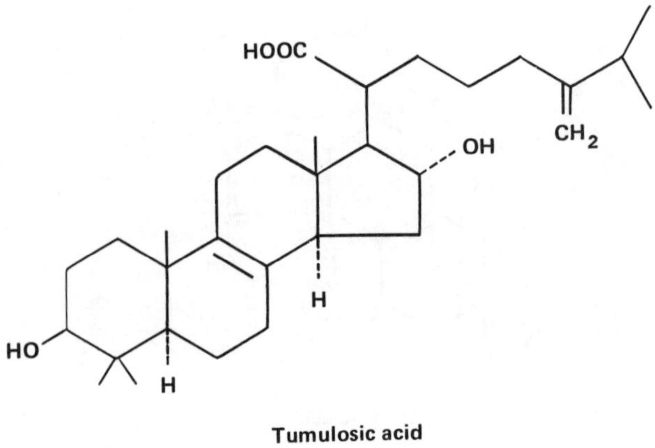

Tumulosic acid

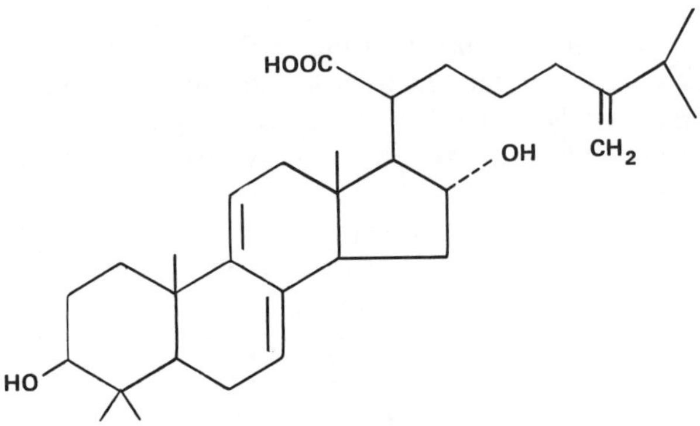

Dehydrotumulosic acid

Polyporenic acid C isolated from *Polyporus betulinus*[40] differs from the other mold acids in having a carbonyl at position 3.

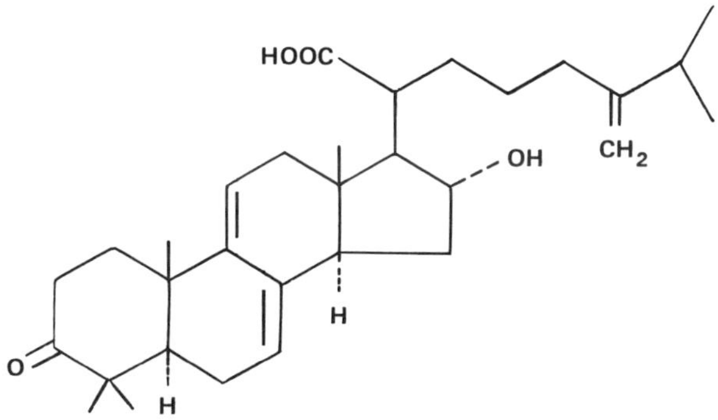

Polyporenic Acid C

It is readily converted to eburicoic acid by chromate oxidation of the methyl ester to yield the 3,16-diketo ester, followed by reduction by sodium borohydride to the 3β-hydroxy-16-keto ester and a Wolff-Kishner reduction of the 16 carbonyl (Figure 2).

Another 4,4,14α-trimethyl sterol acid isolated from *Polyporus* is pinicolic acid A, named after *Polyporus pinicola.* Accompanying this sterol is the 3α-hydroxy acid.[41,42]

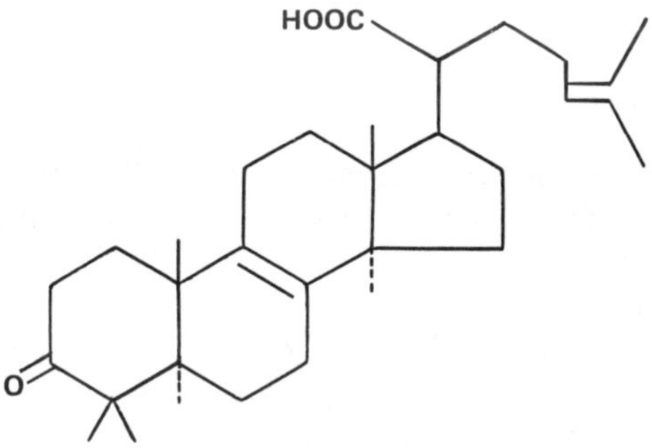

Pinicolic Acid A

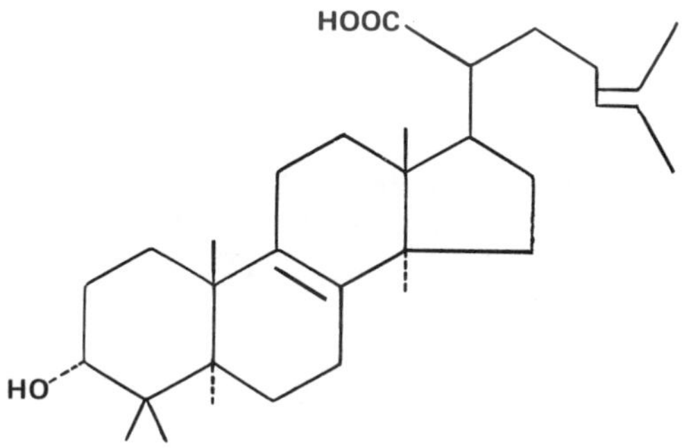

Dihydropinicolic Acid A

A number of 4,4,14β-trimethyl sterol acids are found in plants. In the Manila elemi resin are found "α"-elemolic acid, "β"-elemonic acid, and their $\Delta^{7,9(11)}$ analogues.[43-46]

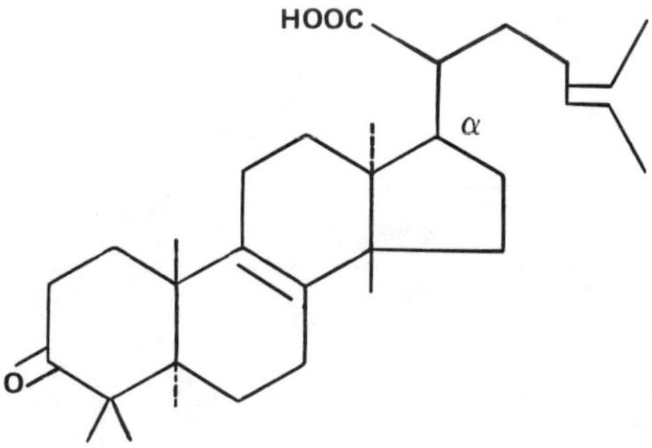

Elemonic acid

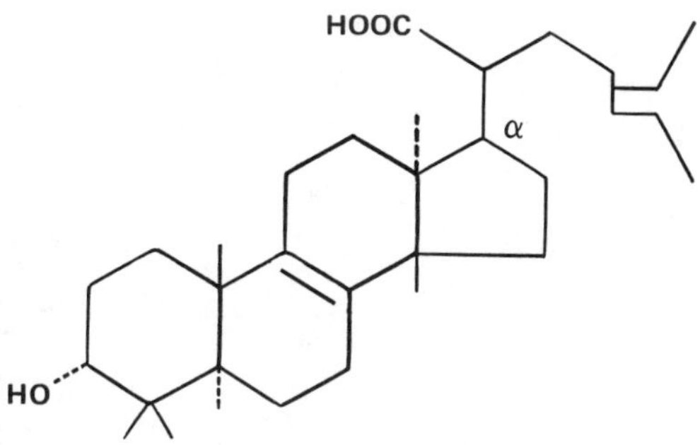

Elemolic acid

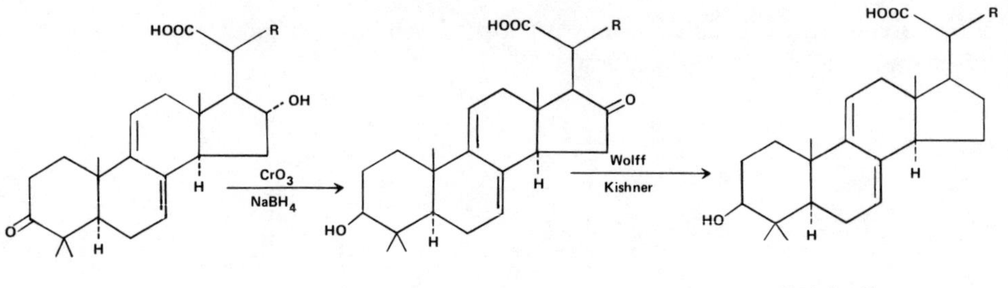

FIGURE 2. Chemical conversion of polyporenic acid C to eburicoic acid.

From gum mastic, a 4,4,14β-acid named masticadienonic acid was isolated.[47,48]

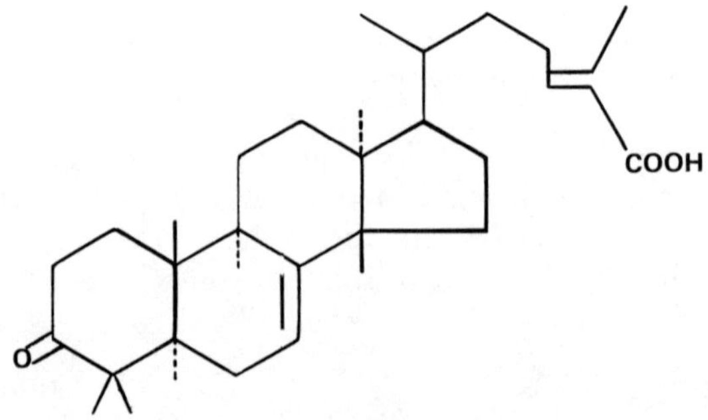

Masticadienonic acid

In the important steroid antibiotics, the acidic function is located at C-21. The presence of the acid group is essential to activity.[49] These include fusidic acid,[50,51] helvolic acid[52,53] and Cepholosporin P.[54,55]

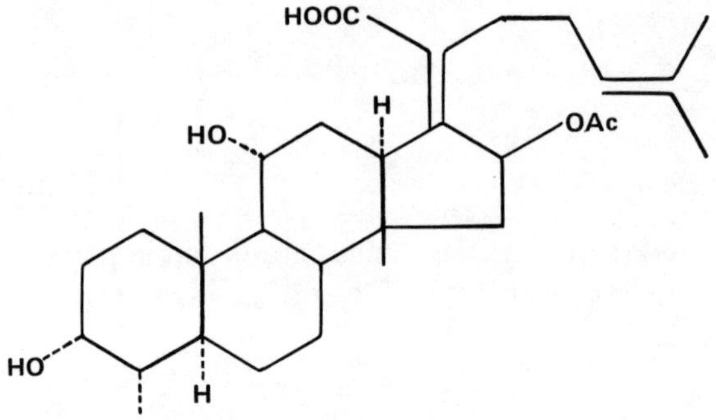

Fusidic acid

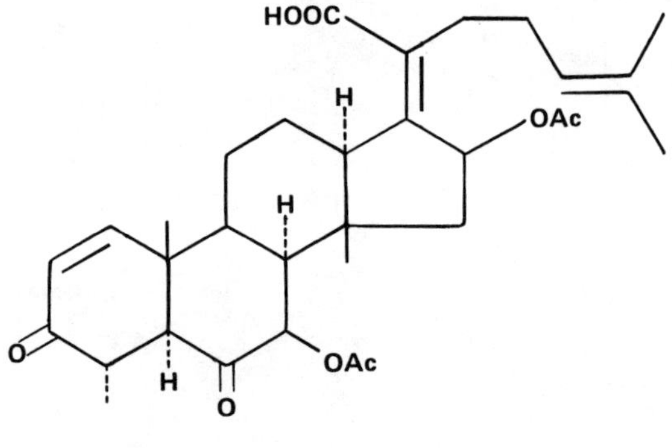

Helvolic acid

TABLE 1

Functional Groups in Coprostanic Acids

Steroid	Hydroxy Groups 3α	7α	12α	C_{27} coprostanic acids: Groups in other positions	Representative source
1	+	+	−	−	Alligator
2	+	+	+	25α-OH	Frog, alligator
3	+	+	+	25β-OH	Frog, alligator
4	+	+	+	23-ene	Toad
5	+	+	+	24-OH	Monitor lizard
6	+	+	+	22-OH	Turtle

IV. THE BILE ACIDS

A. Phylogeny

The bile acids are end products of cholesterol metabolism and are by far the most abundant of the steroid acids. These substances are found throughout the animal kingdom, from fish to mammals. The more primitive organisms synthesize C_{27} acids, whereas the mammals synthesize C_{24} acids, having developed in the course of evolution an ability to cleave carbons 25 to 27 as a propionic acid unit.[56] These C_{24} acids, the cholic acids, represent, then, the contemporary terminus of an evolutionary series of events. Although C_{27} acids have been detected in mammalian bile, they are at best of minor quantitative importance.

The C_{27} coprostanic acids are the end products of cholesterol metabolism in amphibia and certain other nonmammalian species. Examples of these acids and some representative sources are listed in Table 1.

An unusual C_{28} acid, trihydroxybufosterocholenic acid or 3α,7α,12α-trihydroxycoprost-22-enoic acid, has been isolated from toad bile.[57] In marine life other than fish, a bile acid has been reported only once. This acid has an unusual inverted C-20 stereochemistry.[58] Although the cholic and cholanic acids are usually considered to be products of animal metabolism, 5β cholanic acid has been reported in legumes.[59] A C_{24} acid (III) has been isolated from the bulbs of *Eucomis autumnalis*.[60] This acid, the structure of which is shown below

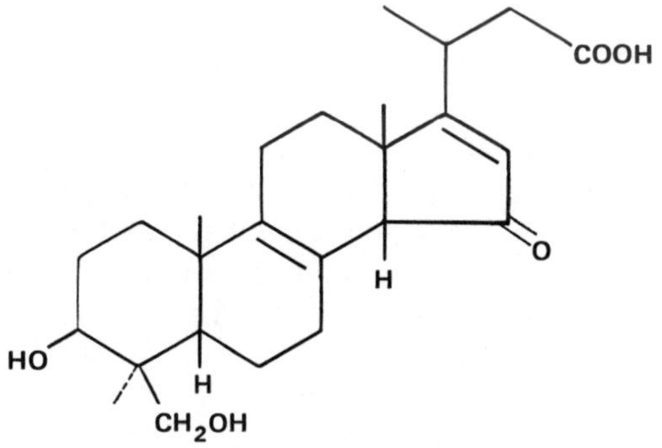

III

is unusual in having the axial carbon at C-4 in a higher oxidation state (see Section III). The variety of C_{24} acids is large. Though certain of them are probably found universally in higher organisms, there is a degree of species specificity. Some of the C_{24} bile acids are products of microbial action in the gut, are reabsorbed by the animal,

TABLE 2

Functional Groups in the Common C_{24} Bile Acids

C_{24} acid	Location of hydroxy groups
Cholic	$3\alpha,7\alpha,12\alpha$
Deoxycholic	$3\alpha,12\alpha$
Chenodeoxycholic	$3\alpha,7\alpha$
Lithocholic	3α
Allocholic (5α)	$3\alpha,7\alpha,12\alpha$

TABLE 3

Functional Groups in the Rarer C_{24} Acids

Sterol C_{24} acid	Location of OH groups	Predominant species
Chenodeoxycholic	$3\alpha,7\alpha$	Birds
Hyodeoxycholic	$3\alpha,6\alpha$	Pig
Hyocholic	$3\alpha,6\alpha,7\alpha$	Pig
α-Muricholic	$3\alpha,6\beta,7\alpha$	Rat
β-Muricholic	$3\alpha,6\beta,7\beta$	Rat
Pythocholic	$3\alpha,6\alpha,16\alpha$	Snake (python)
Bitocholic	$3\alpha,12\alpha,23$	Snake (viper)
Phocaecholic	$3\alpha,7\alpha,23$	Seal, walrus
Tetrahydroxycholanic	$3\alpha,7\alpha,12\alpha,23$	Seal, snake
Allodeoxycholic (5α)	$3\alpha,12\alpha$	Rabbit
Allochenodeoxycholic (5α)	$3\alpha,7\alpha$	Giant salamander
Ursodeoxycholic	$3\alpha,7\beta$	Bear

and are further modified by enzymatic action in the liver before being once more excreted in the bile. This enterohepatic circulation is a common feature of all higher animals. Those acids that are more generally distributed include those listed in Table 2. More specialized biliary distribution occurs with the acids listed in Table 3.

Is is probable that all these acids originate from cholesterol after prior modification of the sterol ring. Some are secondary products of microbial action in the gut, which are reabsorbed and modified still further in the liver. The C_{24} bile acids are by far the major products of cholesterol metabolism and comprise 80 to 90% of the products of cholesterol degradation in humans.[61] On the average, a normal human converts about 1000 mg of cholesterol to bile acids each day. Hepatic cholesterol degradation, accompanied by enterohepatic circulation, controls a major proportion of cholesterol metabolism.[56,62] Because of the efficient recirculation of bile acids, their half-lives are long. Cholic acid has a half-life of 2.3 days and a pool size of 1.4 g, compared to a total pool of bile acids in man of 3 to 5 g, and an enterohepatic circulation of 20 to 30 g daily.[63] The size of the enterohepatic circulation regulates the extent of bile acid formation inversely. The greater the recirculation, the less the production of new bile acids.

B. Oxidation of the Sterol Side Chain in the Synthesis of Bile Acids

Preceding the transformations of the cholesterol side chain in mammals are a series of ring hydroxylations. In the course of cholic acid synthesis, cholesterol is first converted to 5β-cholestane-$3\alpha,7\alpha,12\alpha$-triol.[64] Following hydroxylation at carbon 26 a series of oxidations leads to trihydroxy cholestanoic acid, which is cleaved to cholic acid and propionyl CoA through a series of reactions that probably first involves the conversion of the cholestanoic acid to the CoA derivative.[65] The sites of the 26 hydroxy-

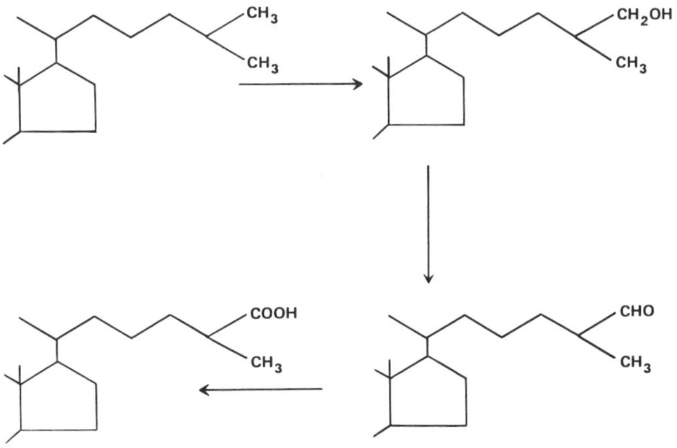

FIGURE 3. Mechanism of formation of C_{27} bile acids.

lation reside in the particulate fraction of the cytosol. Both mitochondria and microsomes have been shown to participate. The modification of the ring system may not be obligatory, since cholesterol itself is hydroxylated in vitro by liver mitochondria[63,66,67] and is transformed to 3β-5-cholenic acid in the rat in vivo.[68] It is not, however, a physiologically significant intermediate on the way to C_{27} or C_{24} acids,[69] since most of the substrate is drained via an initial 7α-hydroxylation. In general, prior hydroxylation of the fused ring system is so rapid that these preferred substrates are available in overwhelming amounts to the oxidizing systems of the liver. The mitochondrial 26-hydroxylase needs NADPH and magnesium ions for the hydroxylation of cholesterol, though the hydroxylation of other C_{27} steroids is not favorably affected by magnesium.[67] 7α-Hydroxycholesterol is hydroxylated at least twice as fast as cholesterol by the mitochondrial enzyme. 7-Hydroxylation may be the rate-limiting step in cholic acid synthesis. It has not been established that the mitochondrial 26 hydroxylase enzyme is cytochrome P-450-dependent, though the requirement of oxygen and inhibition by carbon monoxide is highly suggestive of a P-450 requirement. This mitochondrial location is unusual, for most cytochrome P-450-dependent hydroxylations in liver are associated with the microsomal fraction.[70,71]

A microsomal 26-hydroxylase is found in liver. It is cytochrome P-450 dependent.[72] Unlike the mitochondrial enzyme, it does not hydroxylate cholesterol. In rat liver, mitochondrial and microsomal hydroxylases are of about the same activity, but in humans the microsomal enzyme is of substantially lower activity than the mitochondrial enzyme,[67] in spite of the lability of the mitochondrial 26-hydroxylase.[65] The mitochondrial hydroxylases of rat and human liver have the same substrate specificity. In both species, 5β-cholestane 3α,7α,12α-triol and 7α-hydroxy-4-cholestene-3-one were hydroxylated more efficiently than cholesterol or 5-cholesten- 3β,7α-diol.[63] Hydroxylations at carbon 24 and 25 have been observed, but the extent to which they participate in side chain cleavage and the synthesis of C_{24} bile acids is problematical.[64,73-75] A number of workers have reported that 25-hydroxylation of 3α,7α,12α-cholestane is a significant pathway for the formation of cholic acids, although 25-hydroxysteroids are not converted to C_{24} acids as efficiently as the 26-hydroxylated cholestanes.[68,76,77] It is unlikely that 25-hydroxylation is important in vivo, at least in man, except in certain abnormal conditions, such as in patients with cerebrotendinous xanthomatosis, where 26-hydroxylase activity is probably deficient.[78] The 26-hydroxylated sterol is oxidized to the carboxylic acid in a two-step reaction which utilizes NAD^+ (Figure 3). Oxidation to the aldehyde is catalyzed by the SS isozyme of liver alcohol dehydrogenase.[79,80] The

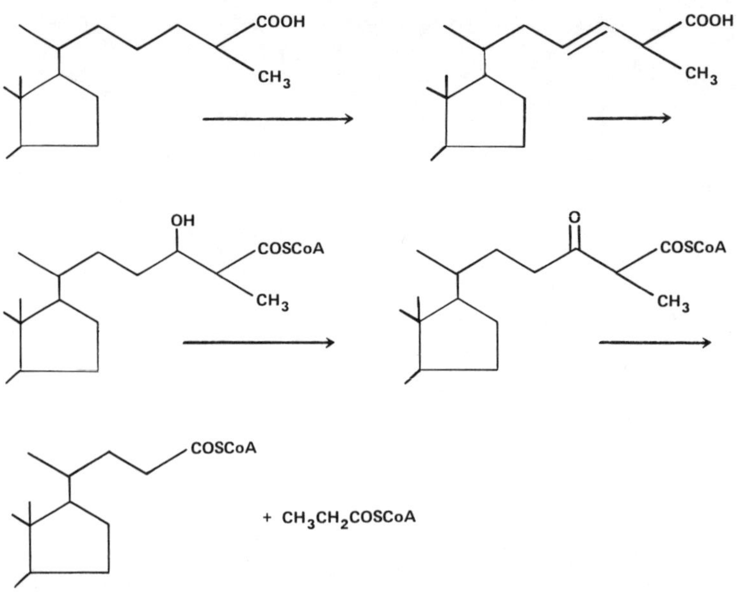

COOH
CH3
COOH
CH3
OH
COSCoA
CH3
O
COSCoA
CH3
COSCoA
+ $CH_3CH_2COSCoA$

FIGURE 4. Mechanism of formation of C_{24} bile acids.

EE enzyme is also active, but to a much smaller extent. The subsequent step, oxidation of 26-al to the 26-oic acid, is mediated by liver aldehyde dehydrogenase.[81,82] Both enzymes are identical to the 26-ol and 26-al oxidases.[83,84]

Cleavage of the side chain results in the loss of a three-carbon unit as propionic acid and the direct appearance of the bile acid. The mechanism may be similar to that of fatty acid oxidation. Thus, 3α,7α,12α-26-oic acid and 3α,7α-dihydroxy-5β-cholestan-26-oic acid may undergo β-oxidation and removal of a propionic acid fragment.[85] These conversions are mediated by mitochondria.[86] The conversion of the 26-oic acid to the C_{24} metabolite may require prior transformation to the CoA derivative[87] (Figure 4). The mechanism requires that a 24-hydroxylated intermediate be formed, but such an intermediate has not been described. The formation of 3α,7α,12α,24-tetrahydroxy-coprostanic acid has been described[88,89] and may be related to the postulated 24-hydroxylation of the 26-oic acid derivative. However, the importance of a 24-hydroxy intermediate has been questioned.[74,87] A mechanism which utilizes the C-24 aldehyde intermediate has been proposed.[90,91] Although it is unlikely that this intermediate is formed, it would be readily converted to cholic acid derivatives because it is a substrate for the aldehyde dehydrogenase normally oxidizing the 26-aldehyde.[90]

V. ACIDIC BACTERIAL DEGRADATION PRODUCTS OF STEROIDS

Total degradation of the steroid nucleus to carbon dioxide and water by bacteria has been firmly established for over 30 years.[92-94] In addition to this complete oxidation, many bacterial strains, especially *Nocardia restrictus, N. opaca, Mycobacterium smegmatis, Glomerella fusaroides, Corynebacterium equi, Arthrobacter simplex, Streptomyces rubescans,* and *S. gelaticus,* are capable of degrading rings A and B to yield 4-(2-carboxylethyl) perhydroindane derivatives. Representative steroids degraded in this manner are summarized in Table 4. The acidic derivative of eburicoic acid is structurally similar to a group of naturally occurring triterpenoid acids, dammarenolic acid,[95] nyctanthic acid,[96] and roburic acid,[97] which have been isolated from plant species. Further metabolism of these acids to a quinoline derivative[98] and to (+)-(5R)-methyl-4-oxo-octane-1,8-dioic acid[99] has been described.

TABLE 4

Degradation of Steroids to Perhydroindan Propionic Acid Derivatives

Precursor	Product	Bacteria	Ref.
Androstenedion	1 Keto-I	Nocardia restrictus	110
Estrone	1 Keto-I	"species	106
Diosgenin	25 R-9 oxide	Nocardia globerula	116
	AB Spirostane,8α		
Hecogenin	propionic acid		116
Eburiocoic acid	3,4-seco$\Delta^{8,24(28)}$ eburicadien-4-o1-3,21 dioic	Glomeralla fusaroides	117
Progesterone	1 Acetyl I	Nocardia sp.	118
Deoxycorticosterone	1(2 hydroxy acetyl)-I	Mycobacterium smegmatis	119
5β-Pregnane-3β,20β-diol	1 Acetyl I	Nocardia sp.	120
	1 (1′ hydroxyl ethyl)I		
Cholic acid	1(4′ valeric acid)	Arthrobacter simplex	121

Note: I = 3(5-oxo-7aβ-methyl, 3aα perhydroindane-4α) propionic acid.

Although several organisms are capable of rupturing ring D to yield a hydroxy acid isolated as testololactone,[100-102] this compound is not further metabolized and is a biological dead end. The opening of ring A in the steroid nucleus was reported by Turfitt,[103] who described the conversion of cholestenone to 5-oxo-A-nor-3,5-secocholestan-8-oic acid by *Proactinomyces erthropolis.* The specific loss of C-4 upon the incubation of 4-^{14}C-cholesterol[104] or 4-^{14}C-testosterone[105] with microorganisms suggested position 4 as a point of attack of microorganisms. The conversion of estrone to a series of metabolites in which ring A was opened to give a series of acids IV, V and VI[106] is in line with this point of view. The first report describing what is

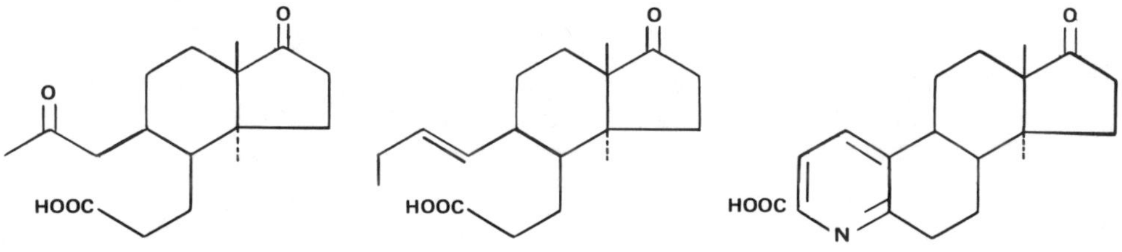

IV

V

VI

most likely the principle pathway of hormone metabolism by bacteria was by Dodson and Muir,[107] who described 9α-hydroxylation followed by 1,2-dehydrogenation and cleavage to give 9, 10-secosteroids. These are further metabolized to give perhydroindane derivatives. The isolation of ^{14}C alanine and L-2-amino-cis-4-hexenoic acid following oxidation with *Pseudomonas testosteroni* is in line with the proposed pathway for the cleavage of rings A and B.[108] Degradation of these perhydroindanes to small chain carboxylic acids has been described by Schubert et al.[109] Following incubation of ^{3}H-3aαH-4α(3′ propionic acid)-7aβmethyl-hexahydroindan-5-one, α-ketoglutaric acid and succinic acid were isolated.

In a careful series of studies, Sih and his colleagues have explored the mechanism by which rings A and B of androstenedione are degraded (Figure 5). The first step[105,107,110] is hydroxylation at C-9 and dehydrogenation at C1,2 in either order to yield 3-hydroxy-9,10-secoandrosta-1,3,5(10)-triene-9,17-dione. This is further hydroxylated to the corresponding 3,4-dihydroxy compound.[111] This is cleaved by a dioxygen-

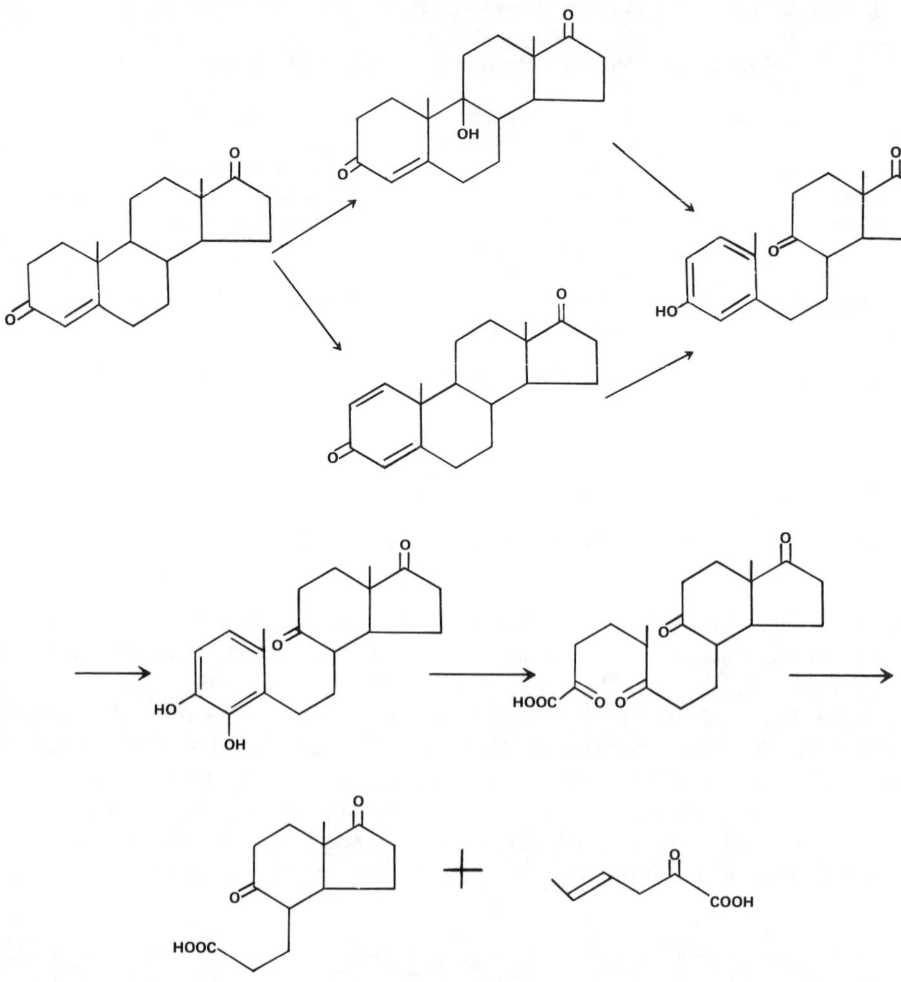

FIGURE 5. Mechanism of bacterial degradation of rings A and B.

ase to yield 4(5),9(10)-diseco-3-hydroxyandrosta(1,10),2-diene 5,9,17-trion-4-oic acid.[112] The cleavage is similar to the cleavage of a variety of catechols.[113,114] This oxidation, as Sih has noted, represents the first positive demonstration of a dioxygenase enzyme in the microbial metabolism of steroids. Ring A is split off as 2-oxo-4-hydroxy caproic acid which is, in turn, enzymatically hydrolyzed to propionaldehyde and pyruvic acid. Rings C and D remain as 3aαH-4α(3′ propionic acid)*7aβ*-methyl hexahydro-1,5-indanenedione (Figure 5). Incubation of estrone yielded a related acid with 4-buten-3-oic acid attached at C-5 (Structure V) because of the failure to further metabolize the aromatic ring A. When either the 1,2-dehydrogenation or 9α-hydroxylation reactions are blocked, further degradation is inhibited.[109] This enzymatic sequence is relatively nonspecific in terms of the functional group at C-17.

Only one step in the enzymatic sequence has been subjected to critical kinetic analysis. Sih et al.[115] showed that oxygenation of 3,4-dihydroxy-9,10-secoandrosta-1,3,5(10)-triene-4,17-dione by *N. restrictus* proceeded by an ordered bi-uni mechanism in which oxygen is added first, followed by substrate and release of product.

VI. ACIDIC METABOLITES OF CORTICOSTEROIDS

A. Historical Introduction

The discovery that corticosteroids are metabolically converted to steroidal carboxylic acids was preceded by a number of observations, which hinted at the possibility that acidic steroids are urinary excretory products. Recognition of these acids was long in coming, because the interest of most students of corticosteroid metabolism centered on the neutral metabolites. To isolate neutral steroids, the organic solvent extract of the hydrolyzed urine was washed with dilute alkali to remove estrogens and interfering acidic substances. It was not generally suspected that among these acid substances were those derived from the corticosteroids. However, the neutral metabolites do not totally account for the radioactivity derived from cortisol when radiotracer doses of steroid are administered to experimental subjects. This has been pointed out by several investigators. Peterson[122] has estimated that neutral metabolites of labeled cortisol after hydrolysis of urine with β-glucuronidase represent about 75% of the total urinary radioactivity and "that 25 percent remains as labeled products not extracted by dichloromethane after β-glucuronidase hydrolysis." Brooks[123] presents evidence for an even greater percentage of cortisol metabolites, ranging from 40 to 55% of the total administered, as unaccounted for as neutral metabolites. Evidence from other laboratories is consistent with these findings.[124-127] In humans, then, a large fraction of cortisol metabolites remain undefined. In the primates, similar conclusions have been drawn.[128] In these, and in other such diverse animals as rat,[129,130] cat, and rabbit,[131,132] the residual unidentified steroids remained in the "alkali backwash" or "could not be extracted from neutral aqueous solution." The properties of these water-soluble metabolites were those of acids. Most workers recognized this possibility and speculated on its likelihood in discussion. In one early study, an acid fraction was obtained from urine of patients with sarcoidosis[133] or a variety of other collagen diseases.[134] The amount was small, representing, at best, 5% of the total steroid excreted.

As a result of in vitro studies and in vivo observations in several laboratories, it became evident that corticosteroids were oxidized to 20-oxo-21-oic acids, 20-hydroxy-21-oic acids, and 17β-carboxylic acids. In the sections to follow, we will review the status of the work on these new metabolites in detail.

B. Nonenzymic Transformations of Corticosteroids to Acids

The lability of corticosteroids to alkali was discovered during the course of studies on the structure of the side chain. The ketol and dihydroxyacetone side chains resemble the terminal groups of the ketol sugars, and like these sugars, it was to be expected that the steroid side chain would be degraded at alkaline pH values, as indeed they were. This was recognized as long ago as 1938, when Mason[135] showed that the dihydroxyacetone side chain of cortisone in 0.04*N* calcium hydroxide underwent changes comparable to those which occur when sugars are treated with alkali. Among the products were etienic and 20-hydroxy acids.[136]

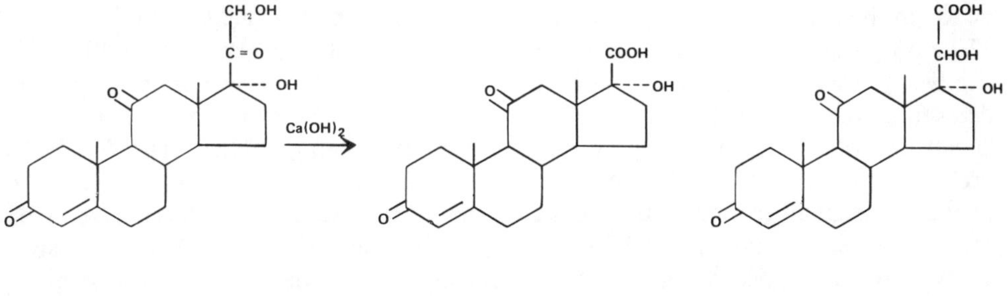

CH$_2$OH | C=O | OH — $\xrightarrow{Ca(OH)_2}$ — COOH | OH — COOH | CHOH | OH

A retroaldol reaction of the postulated intermediate aldotriose may account for the formation of C_{19}-17-ketosteroids,[137,138] which are frequently found as degradation products.

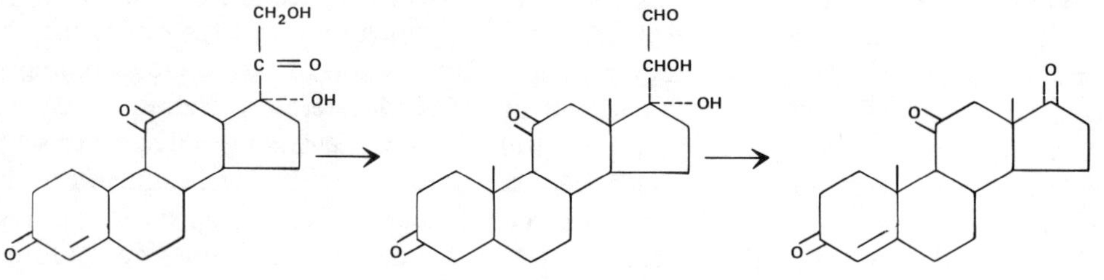

The formation of 20-hydroxy acid could be accounted for by a saccharinic-acid-type transformation of a keto aldehyde, generated by the presence of trace amounts of cupric ion in the water.[139,140]

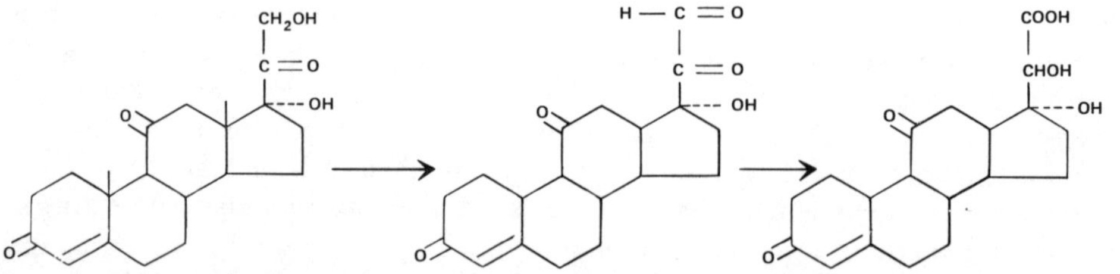

The degradation of corticosteroids to acidic products occurs in homogeneous alkaline solution[141-144] or on the surfaces of adsorbents such as magnesium triacetate or magnesium oxide.[145] These degradations are the consequence of saponification of steroid esters at elevated temperature and are retarded, but not eliminated, when nitrogen replaces oxygen.[143,144,146] The steroid is labile even at neutral pH if held at 15 to 35° for several days.[147,148] This is of particular concern where prolonged equilibrium dialyses are being performed, for the binding of some of the acidic degradation products to serum albumin is greater than that of the parent steroid itself, resulting in increased apparent affinity constants. Careful evaluation of the steroid at the end of any prolonged procedure is essential, because of the possibility of breakdown of the very sensitive side chain.[139,149] When, however, saponification is performed in alcohol under nitrogen with mildly alkaline conditions (e.g., potassium bicarbonate), hydrolyses of C-21 esters occur smoothly overnight[144-146] and the side chain is preserved. Most authors have found that the products of alkaline degradation are the etienic acids and C_{19}-17-ketones.[138,139]

C. The C_{20} Carboxylic Acids

1. Chemical Synthesis of the C_{20} Acids

Corticosteroids are readily oxidized by a number of reagents to C_{20} acids, which are collectively called etienic acids, or in the case of steroids saturated in ring A, etianic acids. In the usual procedure for the synthesis of 17-carboxylic acids, 17-hydroxy and 17-deoxy corticosteroids are cleaved with periodic acid. This reaction is usually quantitative. The acids are readily crystallized and provide suitable derivatives for structural determination. They were made for this purpose very early in the history of steroid chemistry, when the structures of steroids were first being studied. The functional groups on the ring system of the 17β-carboxy acids can be readily interconverted to facilitate the identification of the parent adrenal hormones. Etienic acids derived from the corticosteroids have proved to be valuable derivatives for the chromatographic analyses of the corticosteroids.[150-152] The etienic acid derived from cortisol is highly fluorescent in sulfuric acid-ethanol, and this property has been used to enhance the sensitivity of fluorometric analyses of cortisol.[153,154] Recently the reactivity of the carboxylic acid function has been used to prepare derivatives for affinity chromatography. A stable affinity gel was prepared by coupling periodic acid oxidized corticosterone with aminosepharose through dicyclohexyl carbodiimide. The preparation was used to obtain pure corticosteroid binding globulin in 50% yield.[155] In other cases, steroidal side chains were cleaved to etienic acids by photolysis of organic nitrites,[156] by directed synthesis from Δ^{14}-16-oxosteroids,[157] oxidative cleavage of the side chain,[158,159] base cleavage of the 20-oxo-21-pyridinium iodide,[160] alkali treatment of the ketol side chain,[161] or persulfate oxidation of allopregnane derivatives.[162]

2. Etienic Acids as Metabolites

The oxidative removal of the hydroxymethyl group at position 21 yields a C_{20} acid. This reaction, though readily achieved by chemical means, is affected to only a small extent biologically. Of the few studies in which etienic acids are mentioned as metabolic end products, most have been performed with isolated tissue and organ systems set up for other reasons. For the most part, etienic acids were minor components of steroid metabolites isolated. The earliest workers who reported the isolation of 17-carboxylic acid metabolites of 11-deoxycorticosterone and aldosterone took pains to demonstrate that the acid was formed by a biological process; yet the possibility that these substances were artifacts always lurked in the background, since the interactions were between tissue slices or perfused organs under nonphysiological conditions with supraphysiological amounts of steroid.

The oxidation of corticosteroids to etienic acids was first reported by Picha et al.,[163] who isolated 3-oxo-androst-4-ene-17β-carboxylic acid after perfusion of 11-deoxycorticosterone through rat liver. This transformation also was shown in bovine adrenal[164] and possibly in the rat adrenal.[165] Schneider[166] proposed that the etienic acids originated from C_{21} oxo acids and showed that guinea pig liver slices incubated with 11-deoxycorticosterone did, in fact, synthesize 20-oxo acids and 20-hydroxy acids.

Other corticosteroids are also converted to etienic acids by tissues. Neher and Wettstein[167,168] isolated two steroids, 11β-hydroxy-3-oxo-androst-4-ene-17β-carboxylic acid and 11β,18-dihydroxy-3-oxo-androst-4-ene-17β-carboxylic acid, from pig adrenals. It is unlikely that either is on the main route of aldosterone synthesis. They may have been artifacts generated during isolation of adrenal steroids, since the susceptibility of the corticosteroid side chain to decomposition makes this entirely possible. In other cases, a metabolic origin of the 18-hydroxy etienic acids seems more likely. A steroid with properties resembling 11β,18-dihydroxy acid was formed when 11-deoxycorticosterone was incubated with quartered rat adrenal glands.[165]

Levy et al.[169] perfused 11-deoxycorticosterone through bovine adrenal glands and isolated 3-oxo-androst-4-ene-17β-carboxylic acid 20 → 18 lactone and its 11β analogue. Evidently the cow adrenal gland is capable of converting 11-deoxycorticosterone to 18-hydroxylated etienic acids, though the acids represented less than 1% of the total perfused steroid.[170] In some species, these etienic acid lactones may be normal products of adrenal secretion. Gontscharow et al.[171] isolated 11β,18-dihydroxy-3-oxo-androst-4-ene-17β-carboxylic acid 20 → 18 lactone from adrenal venous blood of the baboon *(Papio hamadryas)*. The sterochemistry of the side chain favors the formation of the cyclic derivatives.

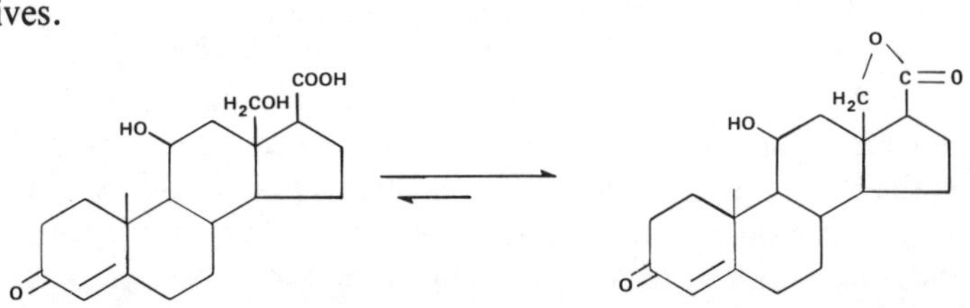

From the studies with animals cited above and our own observations with patients given (1,2-^{3}H)-11-deoxycorticosterone or (1,2-^{3}H)-cortisol,[172] it may be concluded that etienic acid are at best quantitatively minor products of metabolism of the corticosteroids.

In contrast, we have found that 21-dehydrocortisol, VII, a derivative of cortisol whose metabolic role is as yet unclear, is converted by human subjects to the corresponding etienic acid (cortienic acid), VIII, with ease.[173] When 21-dehydrocortisol was given intravenously to human subjects, 11 to 12% of the dose was converted to cortienic acid. This exceeded by far the conversion to any other metabolite. It is interesting to note that of all the steroids recovered, cortienic acid was the only one not reduced in ring A. This suggests that the side chain metabolism of the corticosteroids may, in part, be determined by the oxidation state of the A ring. This point will be considered in Section VI. E.

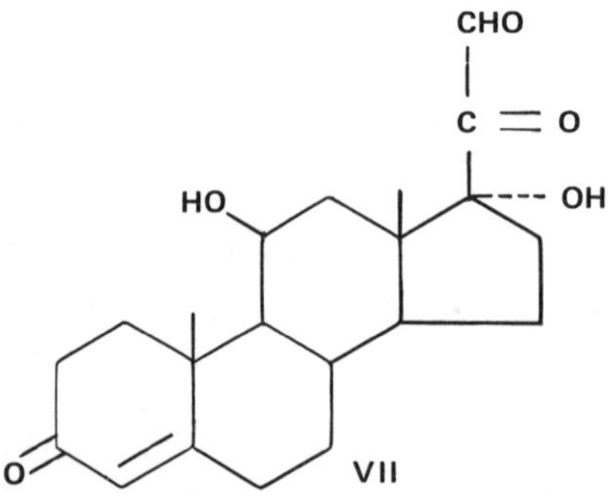

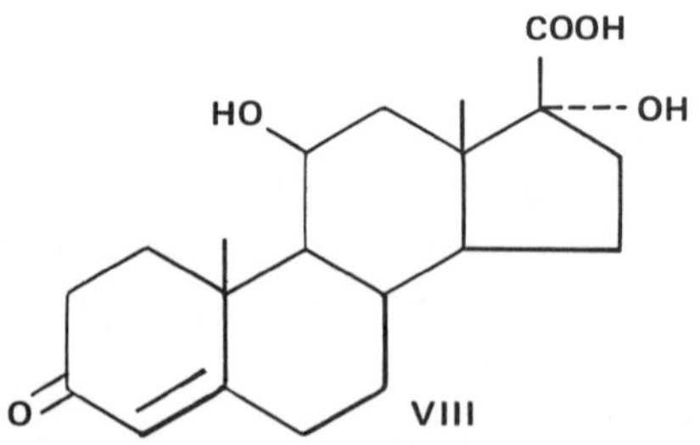

Little is yet known of the metabolic fate of the etienic acids. Benes et al.[174] reported that when 7α-^{3}H-5-etienic acid was administered to a 22-year-old female subject, 32% was recovered in the urine unchanged and unconjugated in 48 hr. The rest was presumably excreted by other pathways. The reduction of 11,17-dihydroxy-3-oxo-androst-4-ene-17β-carboxylic acid to 3β,11β,17-trihydroxy-5-androstane-17β-carboxylic acid was shown by us with rat liver slices.[175] This appeared to be the sole metabolic change. There was no detectable decarboxylation to androgens.

3. *Biological Effects of the Etienic Acids*

The metabolic formation of etienic acids, as we have seen, is quantitatively insignificant. However, they are significant as pharmacological agents. Various etienates have, in fact, been found to inhibit enzyme activity and to interfere with the action of steroid hormones.

Belovsky et al.[176] studied a number of inhibitors of glucose-6-phosphate dehydrogenase, in order to develop an effective treatment for psoriasis. Dehydroepiandrosterone, though a potent inhibitor of the enzyme in vitro, was metabolized too fast by 17β-hydroxysteroid dehydrogenase of red blood cells (the activity of which is increased in this disease) to be effective in vivo. The ethyl, butyl, and octyl esters of 5-etienic acid were effective inhibitors in vitro, though the free acid was not. In vivo, however, the esters were rapidly hydrolyzed and consequently inactive.[174] The 5α-reductases of immature rat testes,[177] skin, [178] sebaceous gland of female hamster,[179-180] prostate,[181] and liver are inhibited competitively by 3-oxo-androst-4-en-17β-carboxylate. The effectiveness of the steroid is consistent with the expected structure deduced by Voigt and Hsia[178] on the basis of studies of structural variations of inhibitory molecules. These results suggest the possibility that the etienic acids may be useful as therapeutic agents in inhibiting the formation of dihydrotestosterone from testosterone, thus suppressing some of the undesirable side effects of testosterone action.

Singer et al.[182] found that the synthesis of cholesterol from acetate was inhibited by 17α-hydroxy-3-oxo-androst-4-ene-17β-carboxylic acid. The Δ^5-3-oxosteroid isomerase of *P. testosteroni,* which catalyzes the isomerization of Δ^5-3-oxosteroids to Δ^4-3-oxosteroids, is competitively inhibited by 3β-hydroxy-androst-5-ene-17β-carboxylic acid.[183]

A number of steroid 17β-carboxylic acids have been tested as part of a program to study the relation of steroid structure to their anesthetic properties. None of the acids, including 3-oxo-5α; 3β,17α-dihydroxy-11-oxo; 3-oxo-5β; 3α,12α-dihydroxy-5β; 3-oxo-4-ene; and 17α-hydroxy-3-oxo-4-ene derivatives, were active.[184] The steroid narcotic 21-hydroxy-5β-pregnan-3,20-dione-21-hemisuccinate Na(Hydroxydione) is metabolized with side chain cleavage to 3-oxo or 3-hydroxy etienic acids, which are inactive.[185]

D. The C-18 Carboxylic Acids

1. Acid Metabolites of Aldosterone

In the key step in the biosynthesis of aldosterone, the 18-methyl group of corticosterone is converted to an aldehyde.[186] The 18-aldehyde is subjected to further oxidation within the adrenal gland. Neher and Wettstein[187] isolated 20,20,21-trihydroxy-3-oxo-pregn-4-en-18-oic acid lactone from bovine adrenals. This acid was a product of the incubation of 18-oxoprogesterone with hog, sheep, rat, rabbit, and bullfrog adrenal slices,[188] along with small amounts of progesterone-18-oic acid. Introduction of an 11-hydroxy group protected against 18-oic acid formation probably by protecting the 18-aldehyde as the 18,11-hemiacetal. Although aldosterone is efficiently synthesized from corticosterone, 21-deoxyaldosterone, and 11-dehydroaldosterone, it is not, by contrast, made from 11-deoxycorticosterone-18-carboxylate.[189] Under physiological conditions, aldosterone-18-oic acid is probably only a minor metabolite of aldosterone. Whether it serves any biological functions at all is uncertain.[188,190]

E. The C-21 Carboxylic Acids

1. The C-20-oxo-21-oic acids

a. Biosynthesis of 20-oxo-21-aldehyde Intermediates

The oxidation of the ketol side chain to oxoaldehyde readily occurs by chemical catalysis (Section VI. B). Despite considerable effort, it has not yet been shown that this oxidation proceeds by enzymatic means. A slow oxidation of corticosteroids to 21-dehydrocorticosteroids is catalyzed by cytochrome c,[191] but the reaction occurs only at pH 8.5 or above with unphysiological concentration of reactants. That net synthesis of steroidal oxoaldehydes occurs under physiological conditions is a reasonable assumption to make. It is difficult to conceive of simple alternative pathways for the conversion of corticosteroids to steroidal oxo acids and etienic acids, and one must infer that an oxidation of the ketol side chain mediated by an alcohol dehydrogenase takes place. A homologous reaction, the oxidation of ethanol to acetaldehyde in the liver, is achieved by three different enzymatic pathways including an NAD^+-dependent alcohol dehydrogenase, catalase, and a microsomal ethanol-oxidizing system.[192] Perhaps the oxidation of the steroidal ketols occurs by comparable routes.

b. The 21-Hydroxysteroid Dehydrogenases

Synthetic 21-dehydrocorticosteroids are readily reduced to corticosteroids by a variety of tissues. The pyridine-nucleotide-dependent enzymes which catalyze this reduction are present in the rat heart, adrenal, liver, kidney, spleen,[193,194] human placenta,[195] and cow adrenal.[196] Three distinct enzymes have been isolated from sheep liver.[197,198] Two are NADH dependent and one is NADPH dependent (Figure 6). These enzymes are highly specific and no other class of substrate has been described. Such restricted specificity suggests that the 21-hydroxysteroid dehydrogenases play a role in corticosteroid metabolism. The enzymes are found in the livers of all species examined throughout the animal kingdom (Table 5).[199] As one rises up the phylogenetic scale, the dehy-

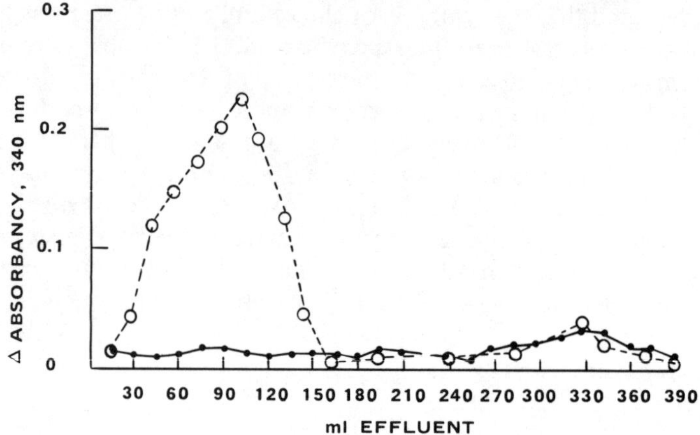

FIGURE 6. Separation of dehydrogenase enzymes. Gel filtration of sheep liver 21-hydroxysteroid dehydrogenases through Sephadex G-100®. Ordinate indicates changes in absorbance per minute at 340 nm produced by a 0.03-mℓ aliquot of effluent in a final volume of 1.0 mℓ. O-----O, NADH-dependent activity; O—O, NADPH-dependent activity.

TABLE 5

21-Hydroxysteroid Dehydrogenase Activities in the Livers of Several Animal Species

	NADH Dependent		NADPH Dependent	
	(μmol Steroid Reduced/100 mg Protein/min)			
	E-I	L-I	E-II	L-II
Mud puppy	1.7	0	0	6.9
Leopard frog	1.2	0	0.8	5.8
Bullfrog	0.7	0	0.8	2.8
Striped bass	0.6	0	0.5	2.2
Yellow pike	0	4.4	0.4	1.1
White Leghorn rooster	0.76	0.6	2.4	3.4
Turkey — male	7.8	0.9	5.7	5.6
Duck — male	9.6	0.5	9.2	5.8
Rabbit — male	7.7	1.5	1.3	3.9
Albino rat — male	2.6	0	0	4.3
Cow — female	30.8	5.5	1.0	3.6
Horse — gelding	8.8	1.3	4.1	2.6
Goat — male	91	3.6	2.6	4.9
Pig — male	1.9	0.7	0.7	4.4
Rhesus monkey — male	2.7	0.4	0.8	6.2

Note: Cytosol preparations were filtered through columns of Sephadex G-200® (1.5 cm × 80 cm). Fractions corresponding to each peak were pooled and analyzed. The designation "0" signifies the absence of a peak at the expected position. During chromatography, pyridine nucleotide dependent activity generally eluted early (E-I, E-II, <65 mℓ) or later (L-I, L-II, >110 mℓ).

drogenase activity tends to increase. Most species contain three or four distinct enzymes. Only necturus and rat contain two: one each of NADH- and NADPH-dependent form.

c. Biosynthesis of the 20-oxo Acids.

The important in vitro experiments of Schneider[166] demonstrated that guinea pig liver slices could oxidize 11-deoxycorticosterone to acidic metabolites. The major product was 3-oxo-4-etienic acid isolated as the glucuronide. Acids represented about 20% of the total steroid recoverd. Schneider proposed that the etienic acid is derived from the α-oxo acid, 3,20-dioxo-pregn-4-en-21-oic acid, which was isolated from the same incubations. Although a precursor-product relationship is supported by this finding, a direct demonstration has not yet been attempted. Subsequently, Gerhards et al.[200] isolated 6α-fluoro-11β-hydroxy-16-methyl-3,20-dioxo-pregn-1,4-dien-21-oic acid from the urine of a patient given fluocortolone. That 20-oxo acids might be products of corticosteroid metabolism appears to be a real possibility.

Progesterone is converted to acidic metabolites by the rabbit.[201-204] These were tentatively identified as C-21-carboxy compounds.[205,206] It is probable that the progesterone is first oxidized by liver to 11-deoxycorticosterone,[207] which is the proximate intermediate. Consistent with this argument, Dey and Senciall[208] found that 11-deoxycorticosterone was converted to uncharacterized acids to a somewhat greater extent than progesterone (41 vs. 37%) by liver fractions.[209] There were substantial species differences in the extent to which progesterone was converted to acids.[210] Rabbit effected the major conversion (66%), followed by guinea pig (24%), pig (23%), rat (19%), and human (1.7%). In the rabbit, two acids derived from progesterone were recovered from urine. Their identities remain in doubt. When esterified and subjected to oxidation by chromic acid, both yielded 3,6,20-trioxo-5α-pregn-21-oic acid-21-methyl ester.[207] It was subsequently shown that the major steroid acid excreted into rabbit urine is 3α, 6α, 20χ, trihydroxy-5β-pregnan-21-oic acid. This was established by comparison with authentic synthetic steroid acid.[213] The authors assigned to this class of metabolite the trivial designation of pregnanoic acid. Some etienic acid also appeared to have been formed. Rabbit liver makes 11-deoxy-corticosterone from progesterone.[211-212] Both steroids are converted to the oxo acid, in accord with the proposal of Schneider,[166] and finally deearboxylated to etienic acid.[213]

Consistent with this possibility is the observation that (21-^{14}C)-progesterone yields $^{14}CO_2$ to an extent equal to 4 to 17% of the dose,[214-217] when administered to mice and rats. Unknown polar or acid transformation products are recovered from the tissues.[214] How corticosteroids are metabolized to 20-oxo acids is unknown. They are probably generated from 21-dehydrocorticosteroid intermediates as depicted in Figure 7. This oxidation has been shown to proceed enzymatically. Oxoaldehyde dehydrogenases, isolated from adrenal[218] and liver,[219] catalyze NAD$^+$-dependent oxidation of steroidal 17-glyoxals to 20-oxo-21-oic acids. The enzymes are located only in organs known to convert 11-deoxycorticosterone to etienic acid. Oxoaldehyde dehydrogenase is an enzyme of broad specificity, which probably serves to oxidize oxoaldehydes generated

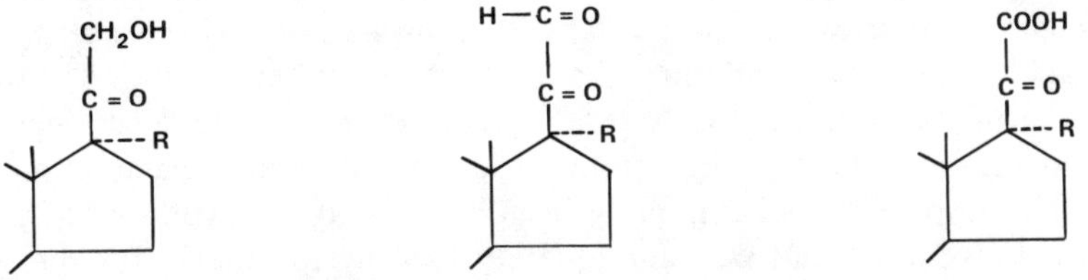

FIGURE 7. Possible route to 20-oxo-acids.

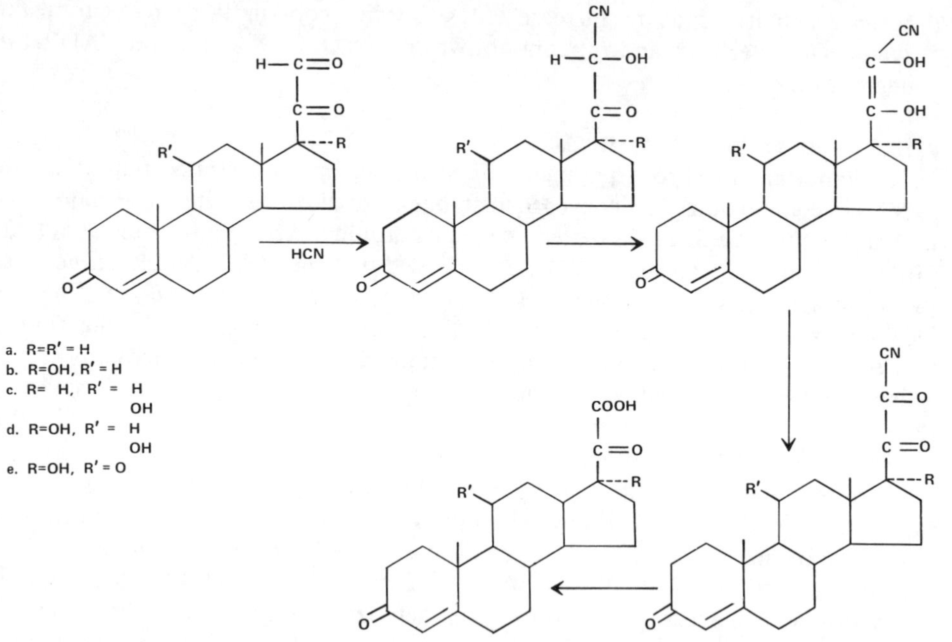

a. R=R' = H
b. R=OH, R' = H
c. R= H, R' = H OH
d. R=OH, R' = H OH
e. R=OH, R' = O

FIGURE 8. Synthesis of 20-oxo-21-oic acids via the cyanohydrin intermediate.

from a variety of precursors, steroidal and nonsteroidal, to oxo acids. The enzyme was originally discovered[220] while studying the metabolic fate of oxoaldehydes generated metabolically from aminoacetone,[221] acetal,[222] and lactaldehyde.[223]

There is an alternative enzymic pathway to oxo acids that does not appear to involve oxoaldehyde dehydrogenase. Enzymes isolated from hamster[224] and human[225] livers catalyze the conversion of 11-deoxycorticosterone to a number of products, among which is 3,20-dioxo-pregn-4-en-oic acid. The mechanism of the transformation is unknown. It is not enhanced by NAD^+, $NADP^+$, or several other hydride acceptors. 21-Dehydrosteroids are not substrates, nor does the formation of oxo acid result from the oxidation of a 20-hydroxy acid. The oxoaldehydes are competitive inhibitors of the enzyme.

d. Laboratory Synthesis of Steroidal Oxo-acids

The 20-oxo-21-oic acids have been synthesized enzymatically in milligram quantities by oxidation of 21-dehydrocorticosteroids with NAD^+ and oxoaldehyde dehydrogenase.[219] The oxidation of steroidal 17-deoxy-oxoaldehydes with silver oxide is an efficient method to prepare 17-deoxy acids. Schneider[166] prepared 3,20-dioxo-pregn-4-en-21-oic acid in this way. The side chains of 17α-hydroxy-20-oxo-21-aldehydes are cleaved off to yield C_{19} steroids under these conditions by alkaline silver nitrate,[226,227] resulting in very poor yields of acids. Excellent yields of 17-hydroxy-20-oxo-acid are obtained by a modification of the method of Schneider.[228]

Another generally applicable procedure is based on the observation that simple α-oxoaldehydes undergo an intermolecular dismutation in the presence of small amounts of cyanide ion to produce a mixture of an α-oxo acid and an α-ketol. The intermediate cyanohydrin may also utilize an external oxidant such as methylene blue.[229] In either case, the oxidation results in the formation of oxo acid and regeneration of cyanide catalyst. This approach had originally been applied to the synthesis of pyruvic acid from methyl glyoxal,[230] but was subsequently used by Monder[231] for the synthesis of steroidal oxo acids from steroidal oxoaldehydes using chromium trioxide or methylene blue to oxidize the cyanohydrin (Figure 8).

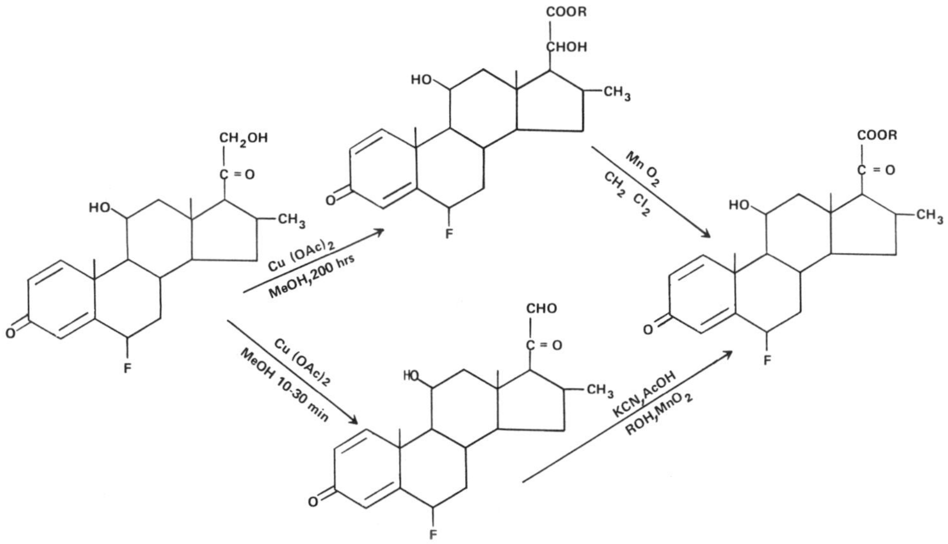

FIGURE 9. Chemical synthesis of 20-oxo-21-oic acid esters.

Several other methods have been devised to synthesize steroidal 20-oxo acids. In each case, in contrast to the procedure described above, the product is the 21-ester of the acid; although readily applied to the synthesis of 17-deoxy oxo acids, it yields 17-hydroxy-20-oxo acids with difficulty, if at all. In the synthesis described by Lewbart and Mattox,[232] the 20-hydroxy acid methyl ester produced by prolonged exposure of the oxoaldehyde or ketol to cupric acetate in methanol is oxidized with chromic acid to oxo acid ester. The chromic acid reagent, however, oxidizes all hydroxyl groups to carbonyl groups. Therefore, the method is of limited applicability.

A modification of Lewbart and Mattox's method was used by Laurent et al.[233] to selectively oxidize 17-deoxy-20-hydroxy-21-oic acid esters to the oxo acid esters (Figure 9). Manganese (IV) dioxide in dichloromethane oxidized the 20-hydroxy group to a carbonyl. Alternatively, the 21-dehydrocorticosteroid stirred with methanol, acetic acid, potassium cyanide, and manganese dioxide rapidly yielded the oxo acid esters by a mechanism similar to that described by Monder.[231] In the procedure used by Laurent et al., solvolysis proceeded with methanol rather than water. A similar approach was used by Corey et al.[234] to convert geraniel to methyl geraniate.

2. *The 20-hydroxy-21-oic Acids*

a. Chemical Studies

i. Synthesis

Schneider, in his classical study on the oxidation of 11-deoxycorticosterone by guinea pig liver slices,[166] isolated 20-hydroxy-3-oxo-pregn-4-en-21-oic acid as a metabolite. This acid was formed in a very small quantity compared to the 20-oxo acid and etienic acid. Interestingly, both 20α and 20β epimers were found. Prior to this observation, Lewbart, Mattox, and Schneider had initiated a thorough examination of the chemistry of steroidal hydroxy acids, which provided a basis for subsequent metabolic studies.

Lewbart entered into this work in an attempt to explain the mechanism of the Porter-Silber reaction[235] and to identify the artifacts caused by traces of cupric ion in the glassware and solution on corticosteroids.[236] The two investigations merged when it was found that both cupric ions and the sulfuric acid component of the Porter-Silber reagent converted cortisone to a common 21-aldehyde intermediate. Extended treat-

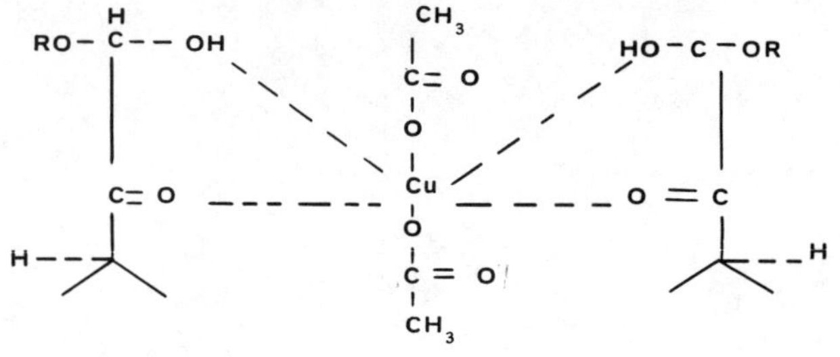

FIGURE 10. Proposed structure of steroid cupric acetate complex.

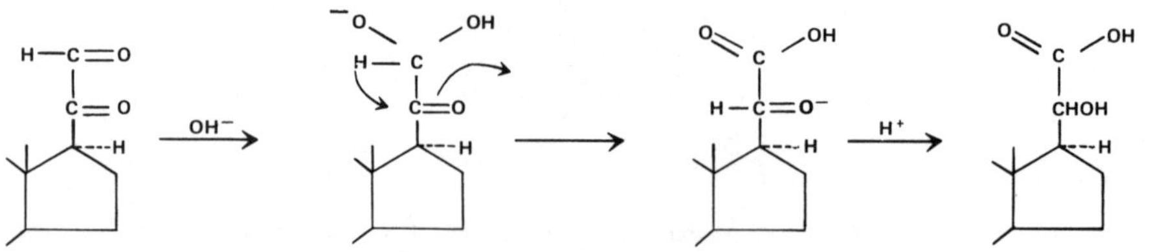

FIGURE 11. Alkali catalyzed rearrangement of ketoaldehydes to hydroxyacids.

ment of the glyoxal with cupric acetate resulted in the formation of acidic artifacts, subsequently identified as the epimeric 20-hydroxy-21-oic acids. Oxoaldehydes are converted into hydroxy acids in two ways. Cupric salts catalyze the side chain rearrangement at a maximal rate when present at a ratio of 1:2 with steroid. Lewbart has proposed that a steroid-cupric acetate complex[236] participates as an obligatory intermediate with the steroidal hemiacetal (Figure 10). A second method for synthesis of 20-hydroxy acids uses sodium hydroxide. Alkali catalyzes a rapid internal dismutation of oxoaldehydes (Figure 11).[237,238]

We have found, in confirmation of the findings of Lewbart, that sodium hydroxide catalyzed the rearrangement of both 17-deoxy and 17-hydroxy oxoaldehydes. The hydroxy acid was formed in 4 to 6 hr from 17-deoxysteroids and 0.5 to 2 hr with 17-hydroxysteroids. The much slower rate of reaction of the 17-deoxy series may be rationalized by the decrease in concentration of the reactive ketol by the formation in alkali of the unreactive tautomeric 17,20-enol.

Cupric acetate converted the 17-deoxy series to 20-hydroxy acids in methanol slowly. The 17-hydroxysteroids gave a poor yield of the desired product. In all cases, the products were found to be 20-hydroxy-21-oic methyl esters.

ii. Chemical Behavior

The presence of the 17-hydroxy group influences the chemical behavior of the hydroxy acid in several ways. Lewbart[239] found that 17,20α-dihydroxy-21-oic acids react with acetic anhydride-pyridine to form 20-acetoxy 17α,21-lactones. The 17,20β-dihydroxy acids, in contrast, yielded 17-hydroxy- 20β-acetoxy-21-oic acid and 17,20β-diacetoxy-21-oic acid, but very little 20β-lactone.

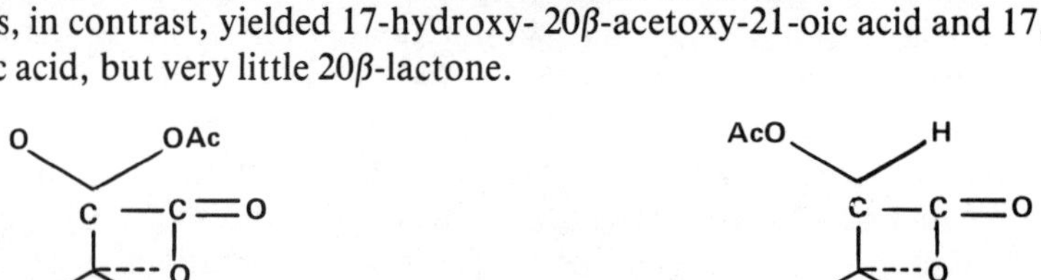

20 α-lactone

20 β - lactone

The limiting factor that determines whether lactone or acetyl formation predominates probably depends on the stereochemical relationships, which facilitate or hinder rearrangement of the intermediate mixed anhydride.

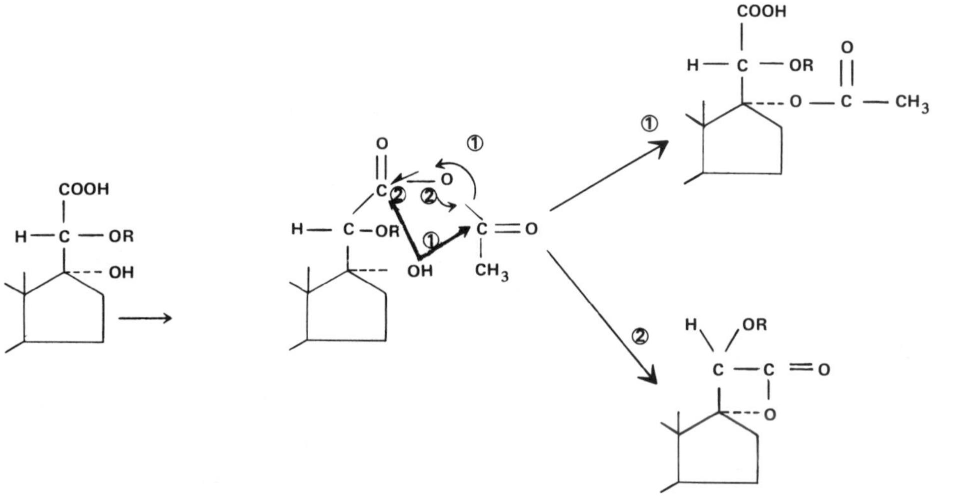

In the 20β series, pathway 2 is presumably hindered because the bulky 20 substituent infringes on C-18 and, by preventing free rotation, blocks the approach of the C-17 oxygen on the anhydride carboxyl carbon. The acetoxy lactones are unstable at room temperature and are decarboxylated to cis and trans enol acetates.

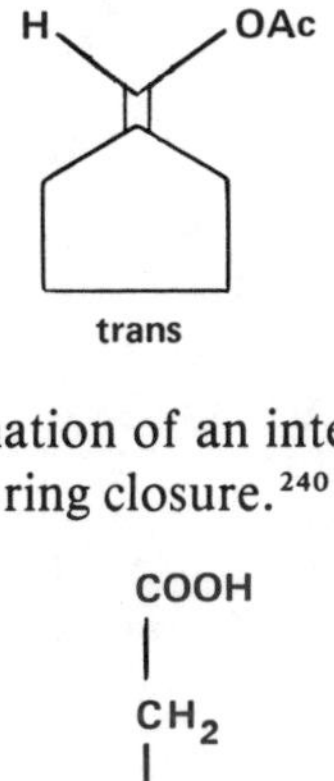

trans

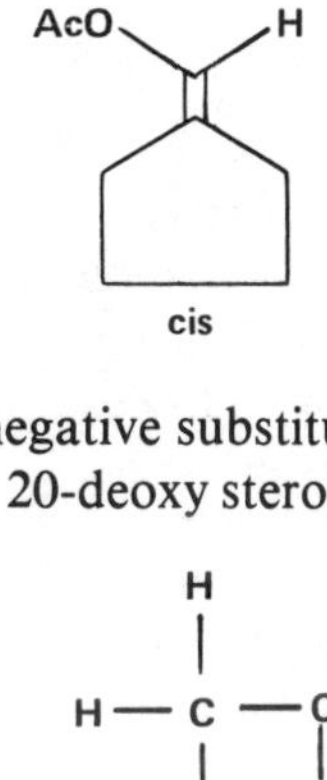

cis

The formation of an internal lactone requires an electronegative substituent at C-20 to facilitate ring closure.[240] Ring closure does not occur with 20-deoxy steroids:

A similar ring closure of the 20α-hydroxy-21-oic acid occurred with ethyl chlorocarbonate

ethyl chlorocarbonate

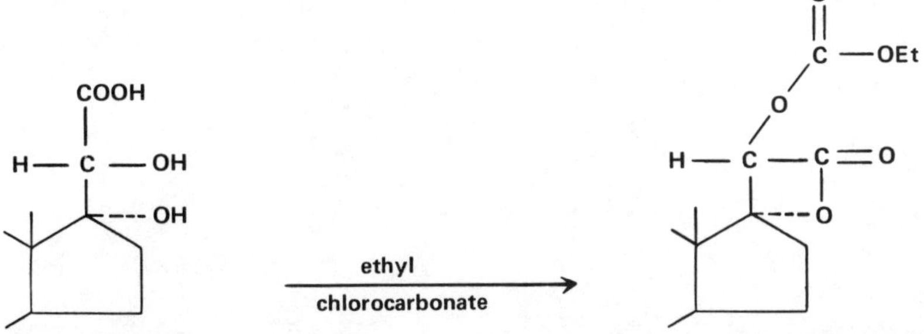

to yield the 20α-cathyl-β-lactone. In methanolic sodium hydroxide, it rearranged to the 20α cyclocarbonate, which epimerized through the 20,21-enol to a mixture of 20α and 20β isomers in which the 20β form predominated.[239]

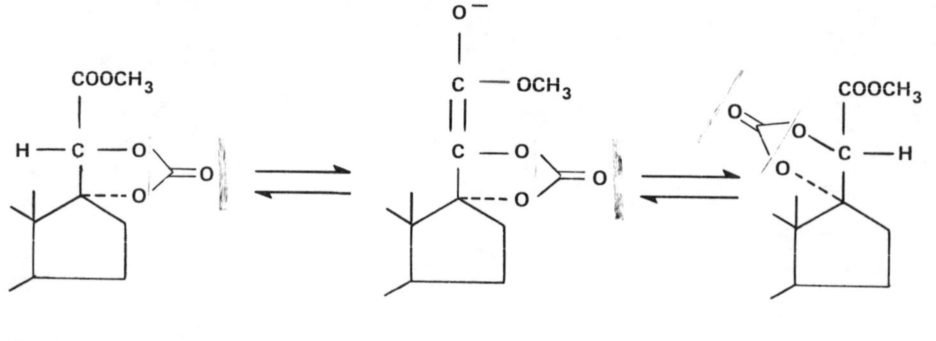

20 α Cyclic Carbonate

20 α Cyclic Carbonate

A similar epimerization is noted at position 20 with 17,20-acetonides of 17,20-dihydroxy-21-oic acid methyl esters.[241] Synthesis of these derivatives occurs by oxidation of the 21-alcohol of the 17,20-acetonide derivatives of the 17,20,21-triols

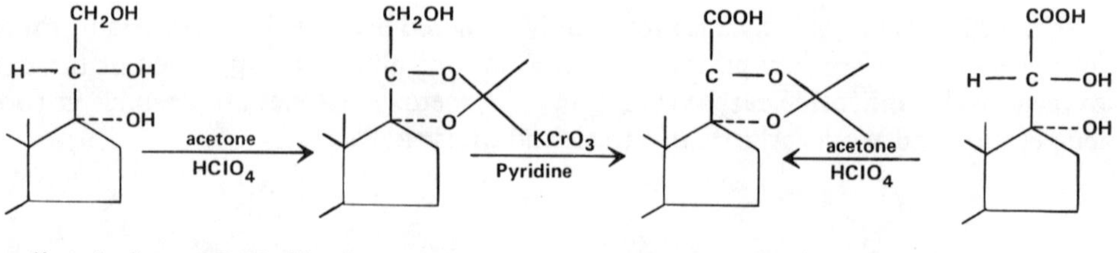

or directly from 17,20-dihydroxy acids.[242] The 17,20α-acetonide 21-methyl ester is converted to a 3:1 mixture of 17,20β- and 17,20α-acetonide-21-oic acids by refluxing with excess alkali. The probable mechanism involves the loss of a proton from C-20 to yield a carbanion. As a consequence, asymmetry at C-20 is abolished by the formation of a resonance hybrid. Readdition of a proton occurs with the formation of the sterically favored 20β acetonide.

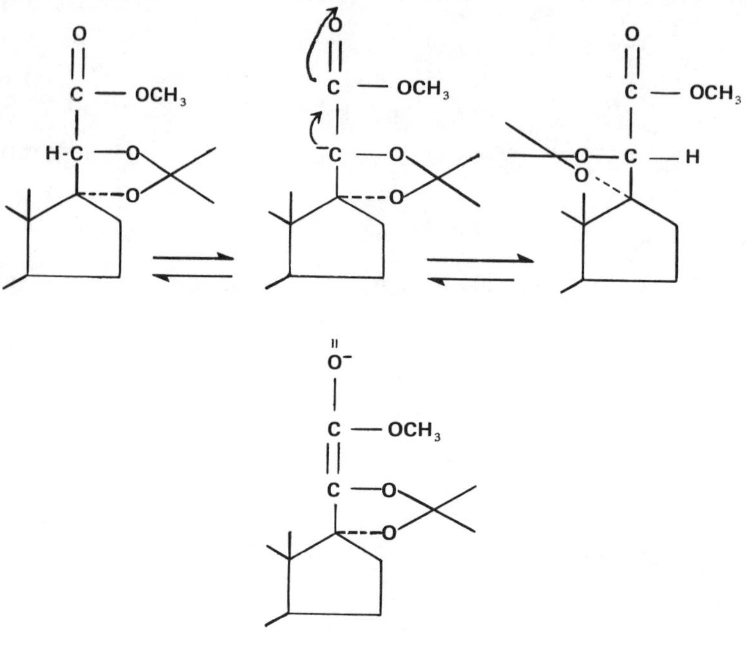

Enolization cannot occur with the free acid, and no epimerization of the acetonide of the 17,20α-dihydroxy-21-oic acid occurs.

In the course of oxidation of the 17,20-acetonide of the 17,20,21-triol to the 21-oic acid by chromic anhydride in pyridine, a 17,20-dioxolone is formed through cleavage between C-20 and C-21 of the 20,21-enol form of the 21-aldehyde intermediate.

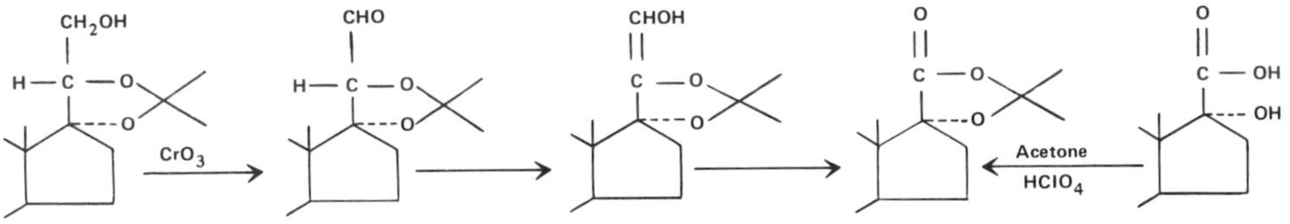

Alternatively, the dioxolone can be formed directly from the 17α-hydroxy etienic acid.[243]

iii. Nuclear Magnetic Resonance

In the above examples of direct epimerization, stereochemical considerations determined the formation of 20β derivatives during rearrangement. That the carboxyl group plays an important role in determining functional group interactions is also shown by NMR analysis. Benn[244] had observed that epimeric Δ^{16}-20 hydroxy steroids could be distinguished by the magnitude of the C-18 shift. In the 20α-hydroxy steroid series, the C-18 signal was downfield to that of 20β. This was confirmed by Robinson and Hofer[245] for other steroids. For methyl 20-hydroxy-3-oxo-pregn-4-en-21-oic acids, acetylation at C-20α and C-20β caused an upfield shift of C-18 of 0.02 and 0.09 ppm, respectively. The effects of acetylation indicated that, in contrast with other steroids studied, side chain conformation of the 21-oic acid series permitted greater deshielding of C-18 by the β isomer.

iv. Optical Rotations

The introduction of a carboxyl group at position 21 produces anomalous optical rotations. The Fieser-Woodward rules[246] state that 20β-acetoxy compounds are more dextrarotatory than 20α-acetoxy derivatives. With 20-hydroxy acid derivatives, these generalizations do not apply.[247-249] In all cases the 20α-acetoxy acids were more dextrarotatory than the 20β-acetoxy series.

b. Biological Studies

i. The Discovery and Isolation of Cortoic Acids in Humans

During the course of studies on the polar urinary metabolites of (4-^{14}C) cortisol, Bradlow and Monder described radioactive water-soluble steroids that were extractable with alkali from ether. These findings are consistent with earlier observations in other laboratories, as shown (Section VI. A). These polar steroids reacted with diazomethane to form relatively nonpolar products. They also reacted with periodic acid to give 11-hydroxy and 11-oxo etiocholanolone, proving that the intact nature of the ring system was preserved. These observations pointed to the presence of a carboxylic acid function in the side chains of these metabolites.

The procedure used in the isolation of the acid metabolites is outlined in Figure 12. The use of Amberlite XAD-2® to isolate these compounds from the alkaline solution proved to be the key step in facilitating their further purification. Figure 13 shows that 85 to 90% of the total steroid extracted by sodium hydroxide consisted of steroids with a carboxylic acid function. Following esterification with diazomethane and further chromatographic separation, a series of carboxylic acid metabolites were isolated. Their structures were established by infrared and mass spectroscopy, elemental analysis, nuclear magnetic resonance, comparison with synthetic compounds with respect

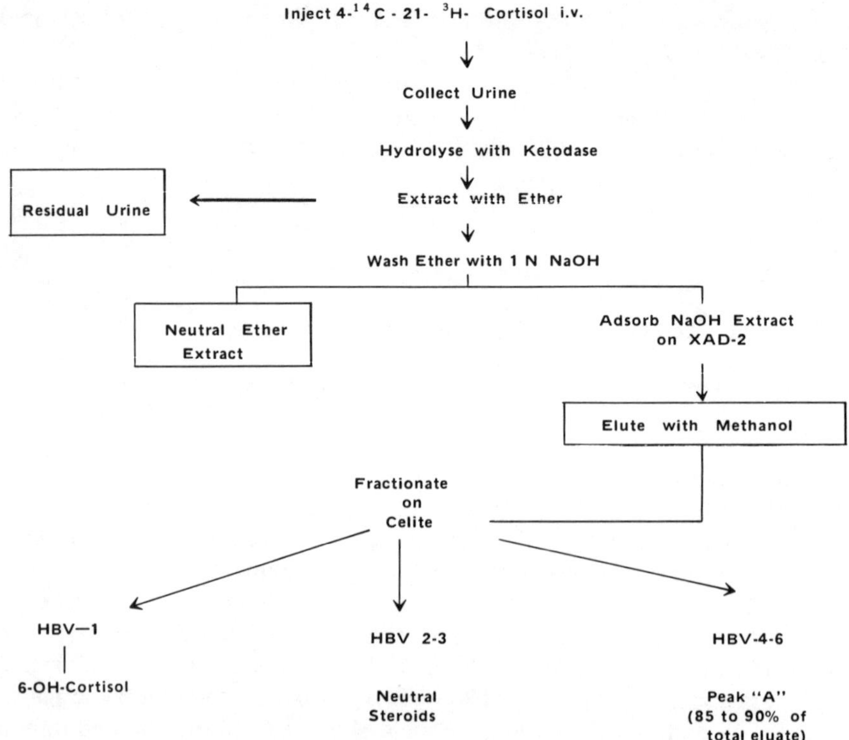

FIGURE 12. Flow scheme for the isolation of acidic metabolites.

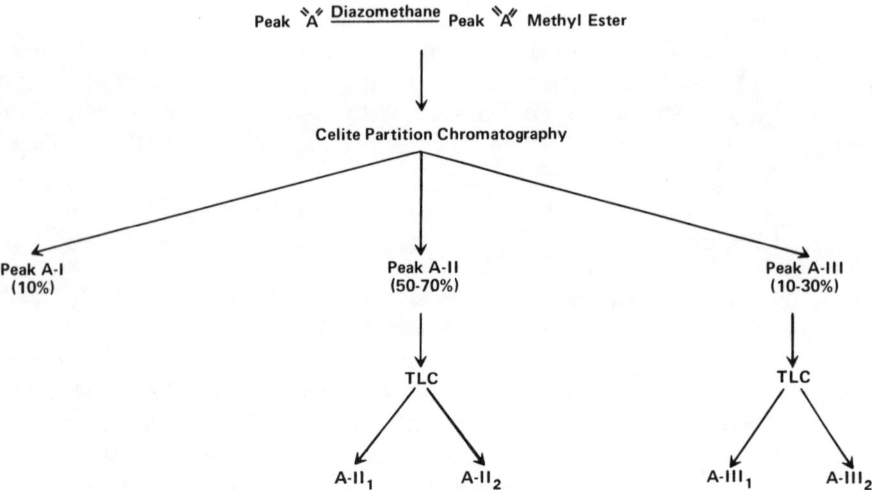

FIGURE 13. Chromatographic separations of cortoic acid esters.

to chromatographic mobility of steroids and derivatives, and reverse isotope dilution.[250] Four were found to be 20-hydroxy-21-oic acids, the structures of which are illustrated in Figure 14. The names of the individual steroid acids are derived from the corresponding 21-hydroxy compounds. We have assigned to these acids the class name of *cortoic acids.* It will be clear to the reader from a comparison of the corticosteroid

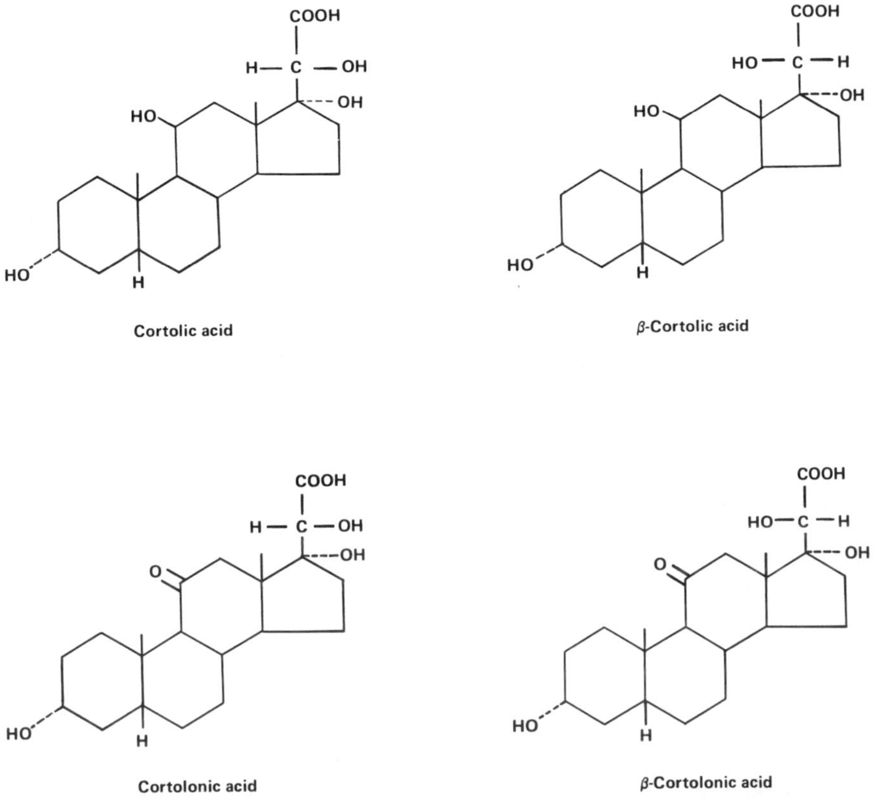

Cortolic acid

β-Cortolic acid

Cortolonic acid

β-Cortolonic acid

FIGURE 14. The cortoic acids.

ketol side chain (IX) and the glycolic acid structure of the cortoic acid side chain (X) that the metabolic conversion of IX to X must involve a complex series of events.

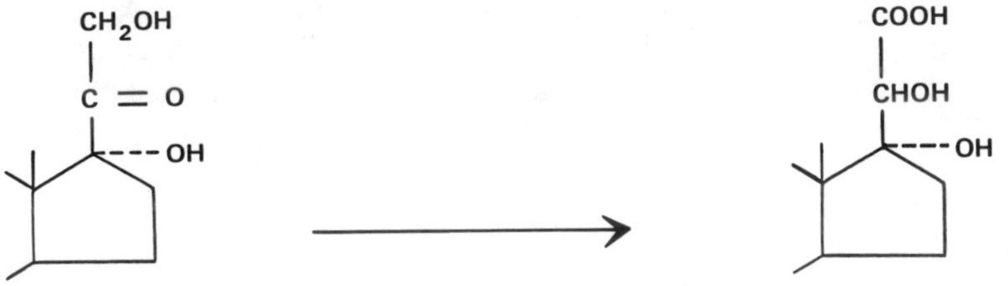

IX → X

ii. Mode of Formation

(a) In Vivo Studies with (21-^{3}H) Cortisol

The oxidation of the hydroxymethyl group occurs with the loss of both 21-methylene hydrogens. Therefore, it was reasoned that if (21-^{3}H)-cortisol could be synthesized and administered to subjects, measurements of the rates of entry of ^{3}H into the body water pool would provide a measure of carboxylic acid formation. Although some tritium did appear in the body water, one of the two tritium atoms at C-21 in the administered steroid relocated at C-20 of the isolated 20-hydroxy acids. This observation suggested that an unusual enzyme-bound transient intermediate, possibly an enediol, was formed.

(b) Studies in Animal Systems

We examined the hamster as a possible model for studying the biosynthesis of steroid

TABLE 6

Effect of Some Tissues In Vitro on Release of Tritium From 21-^{3}H Cortisol and 21-^{3}H-DOC

Tissue	21-^{3}H-cortisol (%)	21-^{3}H-DOC (%)
Liver	5.65	21.43
Kidney	1.18	11.38
Adrenal	0.34	1.69
Brain	0.20	6.29

Note: Steroids (10 nmol, 0.5 Ci) in 25μl of ethanol were incubated for 2 hr at 37°C with 2 ml of 12,000 × g supernatant fraction of 25% (w/v) tissue homogenates prepared in 0.01 *M* potassium phosphate buffer. Reaction was stopped by freezing. Values represent percent of total added radioactivity converted to water.

TABLE 7

Comparison of Substrate Specificity of Human and Hamster Enzyme Complexes

	Activity	
Steroid	Hamster	Human
11-Deoxycorticosterone	100	100
2α-Methylcortisone	14.6	5.7
Corticosterone	11.6	9.3
11-Deoxycortisol	11.5	11.2
Epicortisol	4.5	3.4
2α-Methylcortisol	2.3	3.4
Cortisol	1.8	6.5
6α-Methylprednisolone	1.2	4.4
6α-Fluoroprednisolone	0.9	0.4
Prednisolone	0.9	7.8
Tetrahydrocortisol	0.4	0.9

acids. Following injection of either (21-^{3}H)-cortisol or (21-^{3}H)-deoxycorticosterone (DOC) into hamsters, ^3H_2O was found to be uniformly distributed in the total tody water.[251] Of the several tissues examined in vitro for their ability to labilize the tritium at C-21, liver was the most active (Table 6). The rate of detritiation was greater with DOC than with cortisol. For this reason and because DOC is structurally the simplest of the steroids, it was chosen for all further model studies. Lee has recently reported the formation of similar 20-hydroxy-21-oic acids from (21-^{3}H) prednisolone.[252]

Fractionation of the liver revealed that the activity was concentrated in the cytosol. An enzyme preparation isolated from the cytosol catalyzed the transfer of tritium from (21-^{3}H) DOC to water and the conversion of this steroid alcohol to a series of 21-oic acids comprising the epimeric 20-hydroxy compounds and the 20-oxo acid.[224,225] The hydroxy acids predominated. When (4-^{14}C-21-^{3}H) DOC was incubated with the enzyme, the hydroxy acids contained tritium at position 20, like their human counterparts.[225]

(c) Properties of the Enzyme Complex Isolated from Hamsters

We consider the enzyme to be a complex because it engages in several apparently independent reactions, including the biosynthesis of both 20α- and 20β-epimers of 20-hydroxy acids and the 20-oxo-21-oic acid. We have established that the 20-oxo and 20-hydroxy acids are not interconvertible and, therefore, must have been made via independent pathways by the enzyme. Its molecular weight is approximately 390,000. The preparation absorbs strongly at 403 nm and fluoresces green at long-wavelength ultraviolet stimulation. Its activity is significantly increased by cobaltous ions and strongly inhibited by *o*-phenanthroline. It seems likely, although not clearly established, that one component of the enzyme complex is a metal or metal-containing prosthetic group.

An enzyme complex with similar biochemical properties but considerably lower molecular weight has also been isolated from human liver.[253] It is curious that the human preparations yield more 20-oxo than 20-hydroxy acids, even though in vivo no 20-oxo acids have yet been found. The substrate specificity of the two enzyme preparations is compared in Table 7.

It is noteworthy that both human and hamster preparations are considerably more

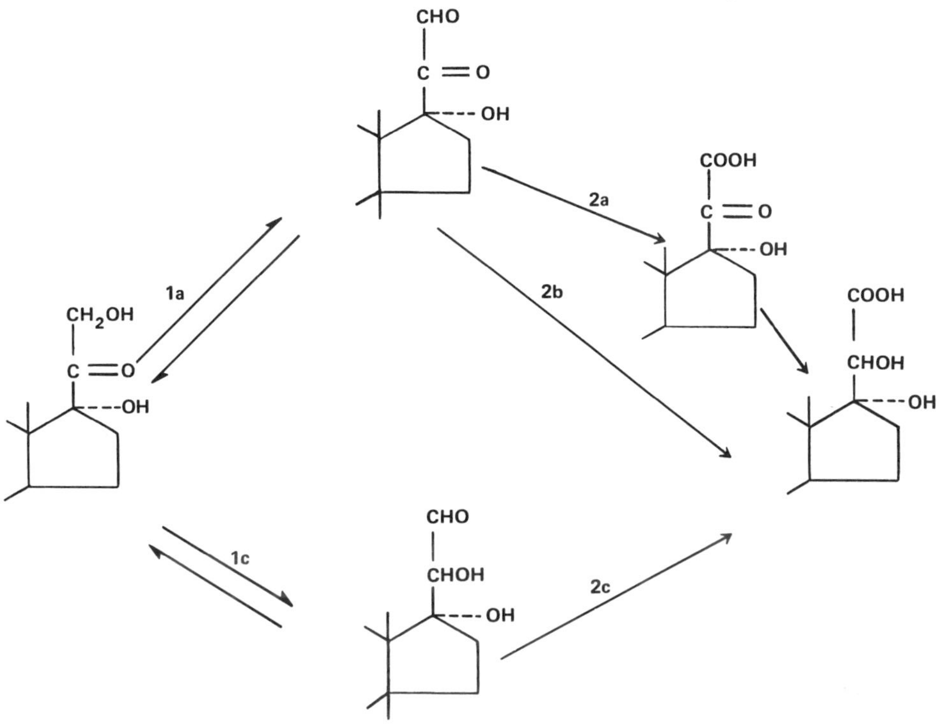

FIGURE 15. Possible pathways from neutral steroids to 20-hydroxy acids.

active with DOC as a substrate than with cortisol. Because the purification of the enzyme was carried out with DOC as the test substrate, it is likely that we have preferentially isolated one form of the enzyme. The greater activity of the crude enzyme preparation toward cortisol[251,253] is consistent with this conclusion.

The formation of oxo acid in greater quantities than hydroxy acids by the purified human liver preparation provides an explanation for the observation of Gerhards et al.,[200] who found that the predominant acidic metabolites of 17-deoxy corticosteroids are 20-oxo acids.

(d) Alternative Metabolite Sequences

The most likely candidates leading to the 20-hydroxy acids are illustrated in Figure 15. Two pathways are distinguishable by the different sequence of oxidation and hydrogen transfer. The pathway through 21-dehydrocortisol is supported in theory by the known ability of alcohols to be oxidized to aldehydes (1A). An aldehyde dehydrogenase that catalyzes the oxidation of oxo aldehydes to 20 oxo acids (2A) was isolated from liver.[219] Reduction of the oxo acid (3A) would then generate the hydroxy acids. This pathway is inconsistent with our findings (see above) that tritium is retained in the side chain. In an alternative pathway utilizing the oxoaldehyde, an internal dismutation[213] would yield the hydroxy acid with retention of tritium directly. To test this pathway, 21-dehydrocortisol was synthesized[196] and administered to patients.[173] Although small amounts of cortoic acids were formed, the overall yield was considerably lower than that from cortisol, whereas if it were in the main pathway, the yield would have been larger. The main acidic metabolite isolated was a 20-carbon acid which we identified as 11β,17α-dihydroxy-3-oxo-androst-4-en-17β-carboxylic acid (cortienic acid). We infer that cortisol may be metabolized via 21-dehydrocortisol to a small extent because cortienic acid was found among the metabolites present in peak A-I

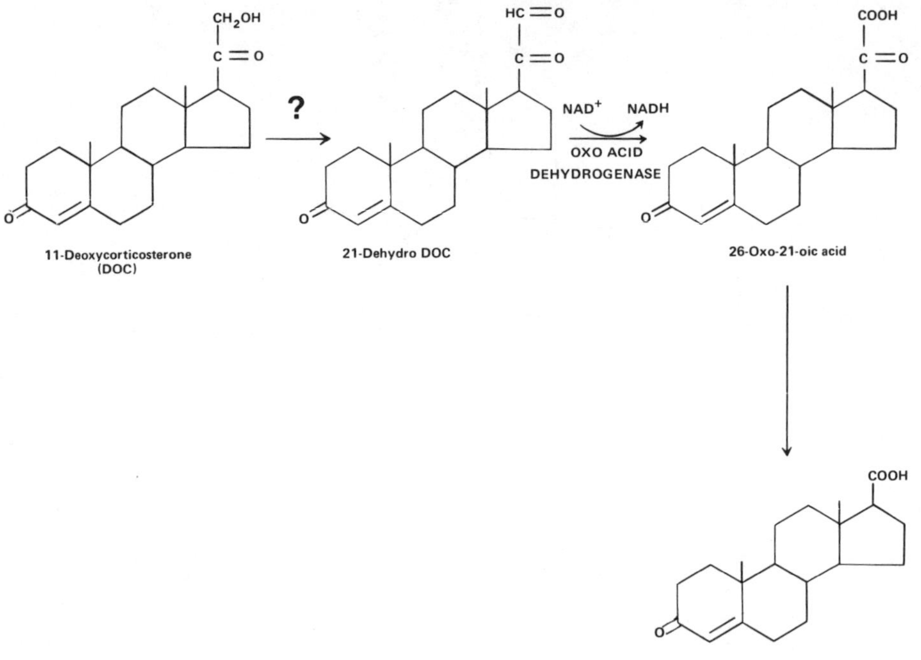

FIGURE 16. Hypothetical route to 20-etienic acid.

(Figure 13) after cortisol administration.[254] The third alternative (1C → 2C) will be discussed below.

(e) Metabolism of 21-Dehydrocortisol in Model Systems

On the basis of in vitro studies with crude hamster liver homogenates or with the purified enzyme complex, there was no evidence for acid formation from 21-dehydrocortisol, although extensive metabolism of the steroid occurred.[224,225] These findings are in agreement with our human in vivo data. It is interesting to note that 21-dehydrocortisol in vivo is not converted back to cortisol,[173] although in vitro reduction of 21-dehydrocortisol occurs rapidly in human and animal liver.

The appearance of cortienic acid in urine after the administration of 21-dehydrocortisol to patients is probably the consequence of an oxidation of the side chain to an oxo acid intermediate, which is subsequently oxidatively decarboxylated to the C-20 acid. This proposed sequence is supported by in vitro observations. Monder[220] isolated an enzyme from mammalian liver which catalyzes the oxidation of oxoaldehydes such as methylglyoxal to the corresponding oxo acids. It was subsequently discovered that steroidal oxoaldehydes are good substrates for this dehydrogenase. This oxoaldehyde dehydrogenase has been found in adrenal cortex[218] as well as liver.[219] In these tissues, 11-deoxycorticosterone has been found to be degraded to the corresponding etienic acid. The enzyme has not been found in other tissues. The findings correlate well with the proposal that the oxidation of oxoaldehydes to oxoacids is an important step in etienic acid formation (Figure 16). Senciall[213] has localized the oxidative decarboxylation in the microsomal fraction of rabbit liver.

(f) Isocortisol as an Intermediate

The alternative route (1C → 2C) in Figure 15 requires the intermediate formation of isosteroids (i.e., steroids with the 20-hydroxy-21 aldehyde side chain) from the parent hormone. The exploration of this pathway was initially handicapped by the una-

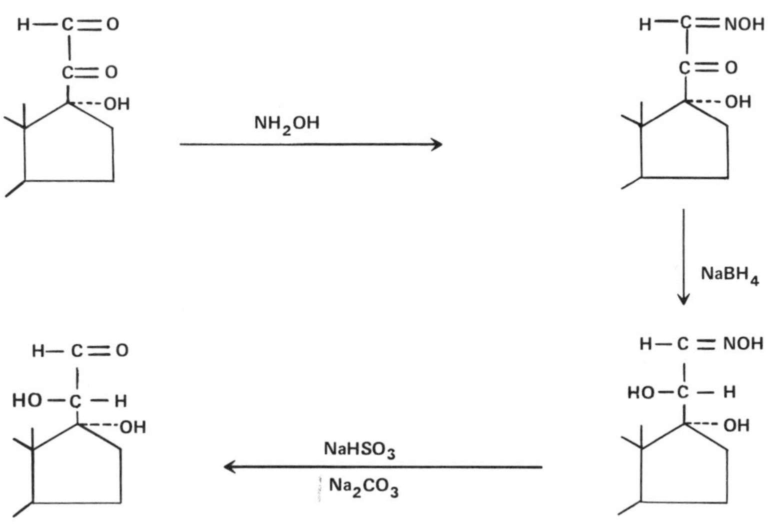

FIGURE 17. Chemical synthesis of 20β-isosteroids.

TABLE 8

Substrate Specificity of 20-Hydroxysteroid Dehydrogenase[a]

Substrate	Rate of reduction (nmol/mℓ/min)	21-Dehydrosteroid corticosteroid
11-Deoxycorticosterone	0.57 ± 0.01	
21-Dehydro-11-deoxycorticosterone	8.51 ± 0.17	14.9
Corticosterone	0.057 ± 0.029	
21-Dehydrocorticosterone	3.86 ± 0.09	67.7
11-Deoxycortisol	1.53 ± 0.00	
21-Dehydro-11-deoxycortisol	6.13 ± 0.25	4.00
Cortisol	0.18 ± 0.14	
21-Dehydrocortisol	2.63 ± 0.02	14.6
Cortisone	3.21 ± 0.02	
21-Dehydrocortisone	5.15 ± 0.10	1.65
Progesterone	1.92 ± 0.01	

[a] The incubation system contained (1.00 mℓ final volume) 90 μmol of phosphate buffer, pH 7.0, 141 nmol of NADH, 289 nmol of substrate, and 1 μg of enzyme. The temperature was 27° ± 1°C.

vailability of the isosteroids. The suggestion that isocorticosteroids are metabolically significant is not original with us. Kendall,[255] as early as 1934, had suggested an aldol configuration for the corticosteroid side chain. This conclusion was based on an erroneous interpretation of the available analytical results. It is curious that after an interval of over 40 years the structure should reappear as a potentially significant metabolic intermediate. Oh and Monder[256] have recently described the successful chemical synthesis of the 20β-epimer as outlined in Figure 17. An alternative enzymic synthesis of the same isomer was achieved by the reduction of the 21-dehydrocorticosteroids with the 20β-hydroxysteroid dehydrogenase of *Streptomyces hydrogenans.*[257] Curiously, the 21-dehydrosteroid is a better substrate of the dehydrogenase than the parent ketols (Table 8). To date, all attempts to synthesize the 20α-epimers have been unsuccessful. The isosteroids are stable when dry. In solution, or even in the presence of small

TABLE 9

Yield of 20-Hydroxy Metabolites from Isosteroids In Vivo

	Isocortisol			Isotetrahy-drocortisone	
Patient	A[a]	B[b]	C[b]	A[c]	D[c]
Neutral metabolites	21.1[d]	24.1	27.0	34.4	35.2
Cortols	8.3	7.7	3.2	2.4	2.3
Cortolones	7.8	10.5	4.8	23.5	20.3
20-Keto metabolites	0	0	0	0	0
C-19 Metabolites	3.4				
Acidic fraction	5.7	8.9	7.5	6.5	3.9
20β-Cortolonic acid	0.2			3.4	1.0
		2.0	2.2		
20α-Cortolonic acid	0.2			0.2	0.1
20β-Cortolic acid	1.1			0.8	0.2
		0.8	1.1		
20α-Cortolic acid	0.3			0.5	—

[a] (1,2^{3}H) Isocortisol + (4-^{14}C) Cortisol® administered.
[b] 4-^{14}C,20α-^{3}H) Isocortisol administered.
[c] (1,2-H^{3}) Isotetrahydrocortisone + 4-^{14}C Cortisol Administered.
[d] Percent administered ^{3}H steroid recovered as metabolite.

amounts of water, these steroids decompose irreversibly. Alkali catalyzes the isomerization of the side chain to the ketol form. 17-Deoxysteroids are isomerized more rapidly than 17-hydroxysteroids, probably because the internal structure of the isosteroid side chain is stabilized by hydrogen binding of the aldehyde to the 17-hydroxy group.[256]

We have administered (1,2^{3}H)-20β-isocortisol or (1,2^{3}H)-20β-isotetrahydrocortisone admixed with (4-^{14}C)-cortisol to human subjects;[258] in other cases (4-^{14}C, 20α-^{3}H) isocortisol was administered. The isolation of 20-hydroxy metabolites as the predominant products isolated after the administration of the isosteroids proves that oxidation or reduction at C-21 proceeds more rapidly than isomerization back to the isomeric ketol (Table 9). This is further supported by the absence of tetrahydrocortisol (THF) and tetrahydrocortisone (THE) as metabolites of the isosteroids.[258] This pattern is quite different from that which occurs after the administration of cortisol or cortisone.[259] Some epimerization of C-20 takes place as evidenced by the formation of α-cortol and α-cortolonic acid. The retention of ^{3}H in the 20α compounds isolated following the administration of the 20α^{3}H, 20βOH-labeled isosteroids suggests a direct epimerization rather than oxidation and reduction at C-20. The absence of the 20-keto metabolites THF and THE further supports this mechanism. The yield of acids from the isosteroids was lower than from the ketols. This was due to rapid transformation of the iso compounds to complexes which remained in the residual urine (Table 10).

Only a small part of the total urinary β-cortolonic acid was recovered from the alkali wash, described in Figure 12, following the intravenous administration of β-cortolonic acid to patients. Modified treatment, including solvent extraction at pH 2 or chromatography on DEAP-LH-20, of the residual urine resulted in increased recovery of the steroid acid. It is likely that these acids are present in a complexed form in urine at neutral pH values (unpublished observations from this laboratory), and that modified techniques will have to be worked out to effect quantitative recoveries of these acids.

(g) Animal Studies with Isosteroids

Two enzymes capable of effecting the oxidation of isosteroids to 20-hydroxy-21-oic

TABLE 10

Percentage of Administered Radioactivity Remaining in the Residual Urine After Enzymatic Hydrolysis of the Urine From Subjects Given Labeled Isosteroids i.v.

Subject	Residual Radioactivity (%)
A[a]	39.1
B[a]	43.2
C[a]	32.4
D[b]	28.7
E[b]	29.5

[a] Isocortisol as precursor.
[b] Isotetrahydrocortisone as precursor.

TABLE 11

Relative Contribution of 21-Dehydrocortisol and Cortisol to Cortisol Metabolites

	Subject	
Metabolite	A	B
Tetrahydrocortisol	0.55[a]	0.57
Tetrahydrocortisone	0.50	0.49
α-Cortol	0.82	0.64
α-Cortolone	8.64	0.55
β-Cortol	1.74	1.69
β-Cortolone	1.43	1.48

[a] Isotope ratio $^3H/^{14}C$ of metabolites divided by the dose isotope ratio (3H-21-dehydrocortisol ^{14}C-cortisol).

acids have been isolated from liver.[260] These are characterized by quite different substrate specificities. Enzyme F-I preferentially oxidizes 17-deoxysteroids. The two enzymes are located in separate compartments of the cell: F-I is in the cytosol and F-II is in the mitochondria. These enzymes are found in human liver[261] as well as in animal liver.[262] It is probable that aldehyde dehydrogenases in human liver participate in the final steps of cortoic acid biosynthesis.

(h) Alternative Pathways in the Biosynthesis of Neutral Metabolites

Both the isosteroids and the 21-dehydrosteroids yield the same major urinary neutral metabolites. These end products are predominantly the hexahydro metabolites (cortols and cortolones). The steps leading from these intermediates which have the 21-aldehyde group in common to the hexahydro end products do not include THF and THE. For example, simultaneous administration of (1,2^{3}H)-21-dehydrocortisol and (4-^{14}C)-cortisol yielded the urinary end products shown in Table 11. It is evident that the cortols and cortolones are enriched in tritium, suggesting that they are formed preferentially from the 21-dehydrocortisol. In order to explain these results, we postulate the existence of an alternative route to the synthesis of hexahydro metabolites which we have called the *long loop pathway.* In this pathway, shown in Figure 18, 21-dehydrocortisol serves as an intermediate precursor of the hexahydro compounds.[173] Reduction at C-20 gives an isosteroid which is in turn reduced at C-21 by an aldehyde reductase to give 20,21-diols. The neutral metabolites derived from the isosteroids are predominantly cortols and cortolones. Remarkably, although only the 20β-isosteroids were given, both epimeric cortolones were isolated (Table 12).

(i) Animal Studies on Isomerization and Epimerization.

The conversion of isocortisol to the 17,20,21-triol is mediated by a 21-aldehyde reductase[263] which can be separated from the NADH-dependent 21-hydroxysteroid dehydrogenases described by Monder and White (see Section VI. E. 1b.). This enzyme, which requires NADPH as a cofactor, has relatively high specificity for the hydroxy aldehyde configuration. It has been isolated from horse, hamster, and rat liver in good yields. Neither gylcolaldehyde nor glyceraldehyde-3-phosphate is a substrate.

Both isoDOC and DOC are converted by crude hamster liver homogenates to 20,21-diols but not to the 20-hydroxy acid. IsoDOC is also rearranged to DOC. The transformation of isoDOC to DOC appears to occur more rapidly than its conversion to

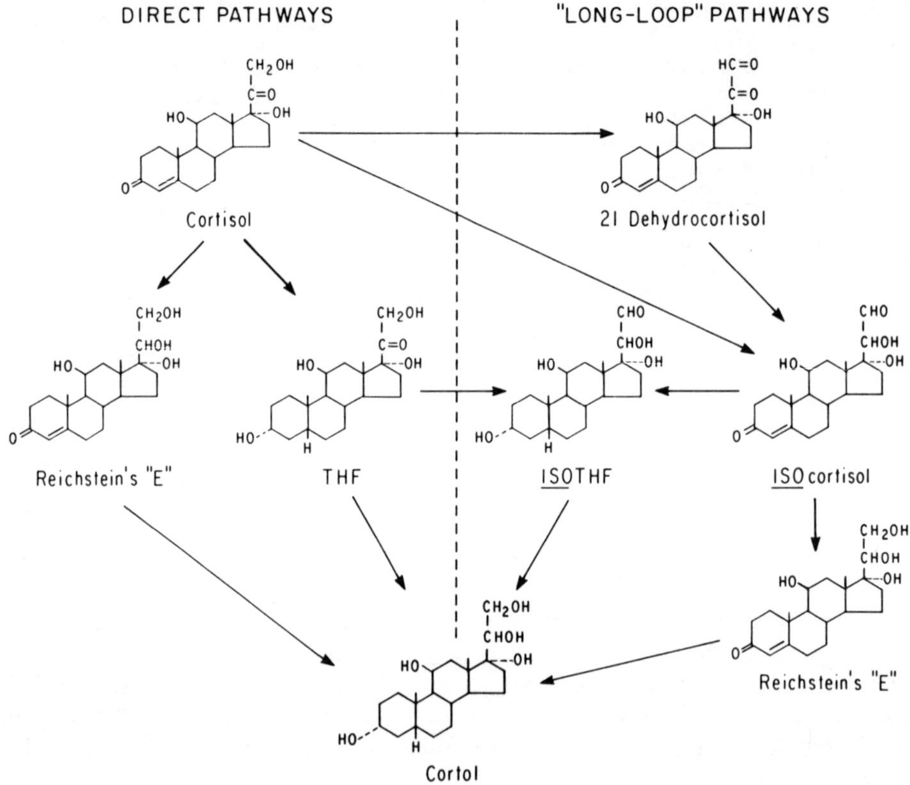

FIGURE 18. Steroid metabolic pathways involving 21-dehydrosteroids.

TABLE 12

Relative Conversion of Isosteroids to Neutral Metabolites[a]

	Relative Isotope ratio				
	Isocortisol			Isotetrahydrocortisone	
Metabolites	Patient A	Patient B	Patient C	Patient A	Patient D
Cortol α	0.71	1.01	—	—	0.63
Cortol β	1.80	1.05	1.01	0.62	0.71
Cortolone α	0.53	0.96	0.86	1.10	1.39
Cortolone β	0.89	1.01	0.87	2.61	2.15
THF	<0.01	<0.01	<0.01	<0.01	<0.01
THE	<0.01	<0.01	<0.01	<0.01	<0.01

Note: Each isosteroid was administered admixed with ^{14}C-cortisol and the $^{3}H/^{14}C$ ratio of each metabolite was divided by the dose ratio.

[a] See Table 9 for detailed treatment of patients.

the diol. This suggests that in crude preparations isoDOC is not reduced to diol directly but rather through DOC. The use of ($20\alpha^{3}H$) isoDOC, however, indicated that in this system two independent reactions were occurring: one to the diol and the other to DOC. Though reduction of DOC took place, direct comparison showed that isoDOC was the preferred substrate leading to the diol.

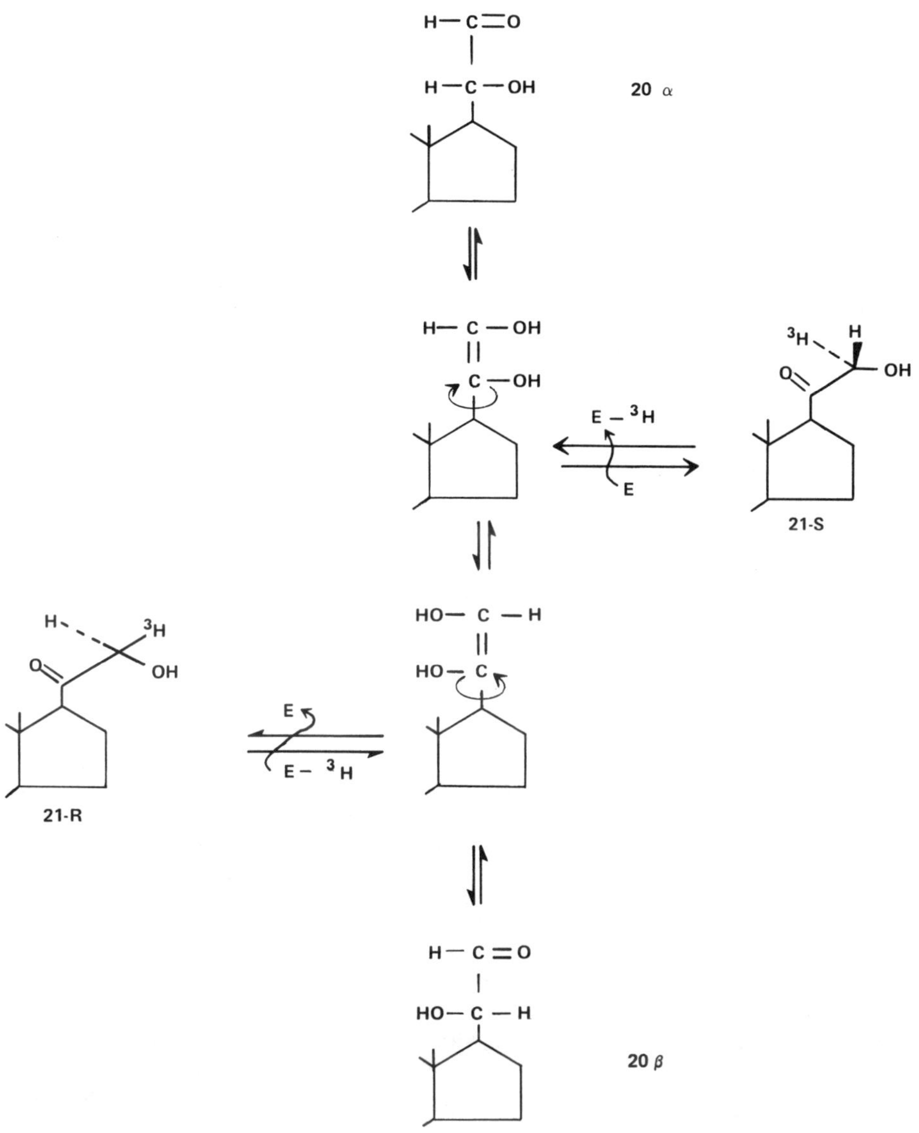

FIGURE 19. Working model for isomeric and epimeric interconversion of 21-^{3}H-DOC and isoDOC.

The isomerization of isoDOC to DOC is catalyzed in hamster liver by a large enzyme complex of molecular weight approximately 400,000.[225] It appears probable that the enzyme catalyzes the reversible interconversion of isoDOC to DOC by a mechanism reminiscent of the interconversions of the ketose and aldose sugars.[264] If the same mechanism of interconversion occurs with corticosteroids as with sugars, then experimental approaches developed from the latter should yield similar results for the former. Starting with this premise, we have developed the working model shown in Figure 19. In accordance with the predictions from the model, incubation of (21-^{3}H)DOC with enzyme led to the exchange of tritium with protium of water. A similar experiment in which DOC was incubated with ^3H_2O in the presence of the enzyme yielded (21-^{3}H)DOC. The incubation of (20α^{3}H)-isoDOC with enzyme led to the formation of (21-^{3}H)DOC and the incorporation of tritium to a measurable extent into water. The results are clearly consistent with the identification of the enzyme as an isomerase which, like the aldose-ketose isomerases, catalyzes the interconversion of the aldol and ketol forms of the side chain through an enediol intermediate.

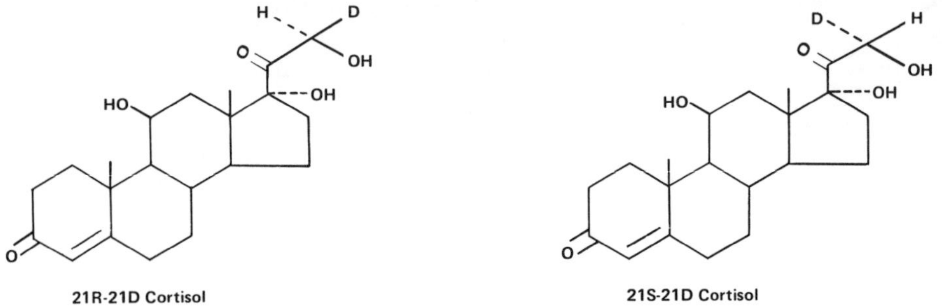

FIGURE 20. Prochiral centers at position 21 of cortisol.

It should be noted from the model that the tritium from C-20 and C-21 is transferred to a basic group in the enzyme during the formation of the common enediol intermediate. This tritium may be returned to the steroid at either site or exchanged with water. Consequently, over long periods of incubation, a cycling process takes place that ultimately results in the complete loss of tritium from the substrate and its appearance in water. This reaction occurs at transfer-exchange ratios which are characteristic of the isomerase in question.[264] The ratio for the steroid ketol-aldol isomerase strongly favors tritium exchange relative to intramolecular transfer.

(j) Stereochemistry of the Steroid Side Chain

It is evident from Figure 19 that corticosteroids are prochiral at position 21, that is, the substitution of one of the methylene hydrogens with deuterium or tritium results in the appearance of an asymmetric center. The asymmetry has been designated in Figure 20 according to the nomenclature of Cahn, et al.[265] as 21R and 21S. We have devised methods for the specific synthesis of each epimeric form (Figure 21)[266,267] in order to study the stereospecificity of the enzymes that catalyze side chain metabolism.

The isomerase-catalyzed loss of tritium from the 21S isomer was much more rapid than that of the 21R epimer consistent with stereoselectivity for the enzyme.[251,260] If absolute stereoselectivity prevailed, no tritium should have been lost from the R epimer. However, loss did occur. One can account for our unexpected results by assuming (1) the presence of two distinct enzymes of unique stereospecificity, (2) one enzyme capable of acting on both epimers at different rates or (3) the prior transformation of 21R to 21S by an epimerase reaction. It has not been possible to demonstrate two separate enzymes. Assumption 1 does not, therefore, explain our results. The action of the same enzyme on two epimers at different rates should follow Michaelis-Menten kinetics for both reactions. We find that simple kinetics is followed for the 21S form. The 21R steroid participates in the reaction only after a lag period. This is inconsistent with Assumption 2. The existence of a lag period suggests that Assumption 3 is a likely explanation. This was tested by measuring the rate of incorporation of deuterium from D_2O into the C-21 position of unlabeled DOC.[260] Initially, a single deuterium was incorporated at C-21. After a lag, a second deuterium appeared in position C-21 leading to dideuteroDOC, as predicted by the epimerase mechanism. The proposed mechanism predicts that the epimerase should also convert 20βisoDOC to the 20α form, as it indeed does. A model incorporating these observations is presented in Figure 19.

Evidence for epimerization in vivo was obtained in the studies on the fate of isosteroids in man. Although only the 20β-isosteroid was administered, both epimers of cortolone were obtained (Table 12). To establish that the reaction did not involve an oxidation-reduction sequence with intermediate formation of the 20-oxo-21-aldehyde, the fate of (4-^{14}C-20α^{3}H) isocortisol was studied. Here, too, the α-cortolone isolated

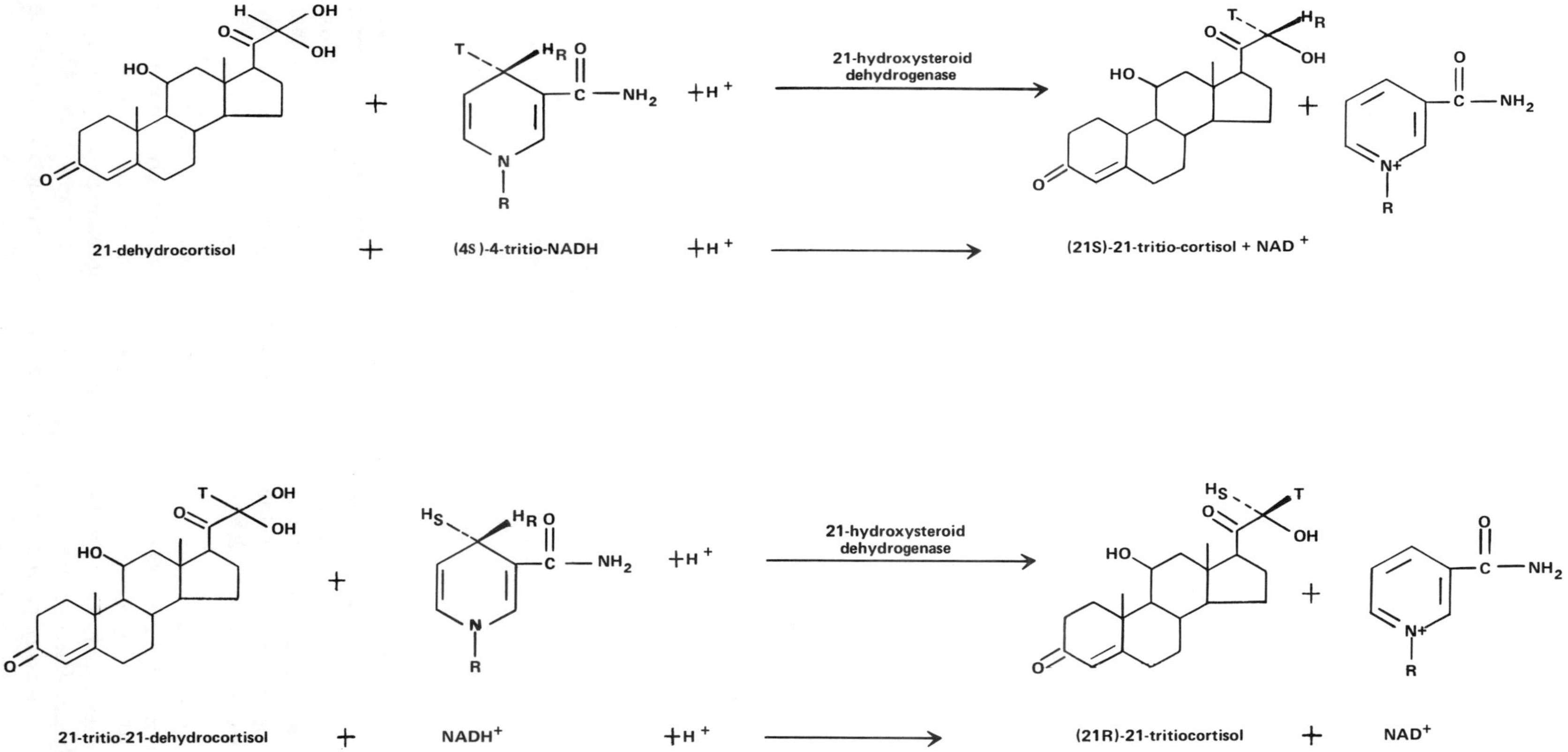

21-dehydrocortisol + (4S)-4-tritio-NADH + H^+ —21-hydroxysteroid dehydrogenase→ (21S)-21-tritio-cortisol + NAD^+

21-tritio-21-dehydrocortisol + $NADH^+$ + H^+ —21-hydroxysteroid dehydrogenase→ (21R)-21-tritiocortisol + NAD^+

FIGURE 21. Synthesis of enantiomeric forms of 21-^{3}H-cortisol.

still contained ^3H, showing that oxidation-reduction of the side chain could not have occurred.

(k) Preferred Pathways of Cortoic Acid Formation

We have calculated that the number of possible permutations of the steps leading from cortisol to cortolonic acid is 12.[268] It is self-evident that not all of the 12 routes are quantitatively important. Comparative studies on the relative yield of cortoic acids from the ring A-reduced metabolites THF and THE revealed that these compounds are better precursors of the acids than the parent hormone and indeed may well be obligate intermediates.[269] THE preferentially yielded the cortolonic acids and THF the cortolic acids, as might have been expected from the parallel pathways leading to the neutral metabolites of these compounds. Figure 22 sums up our current understanding of the probable pathways through which cortisol is metabolized to neutral and acidic metabolites.

(l) 17-Deoxy Acidic Metabolites

It has been universally accepted that during the metabolism of 17-hydroxylated corticosteroids the 17-hydroxy group remains intact. From fraction A-I (Figure 13), two steroids were recovered that did not correspond in properties with the known cortoic acids. These steroids were identified by us as 17-deoxy cortolonic acids distinguished by

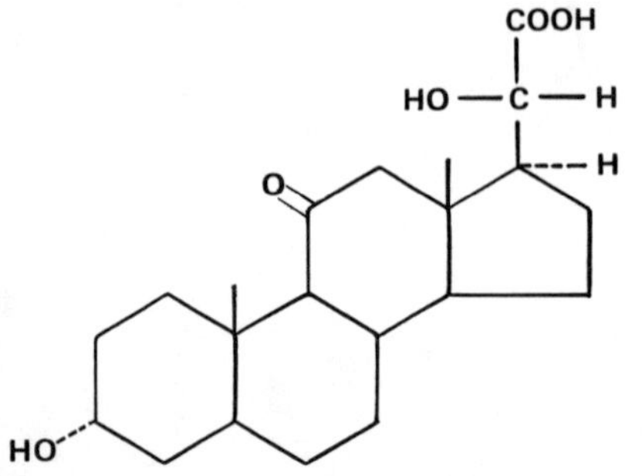

COOH | HO—C—H | ---H ... HO

20 α - Deoxycortolonic acid

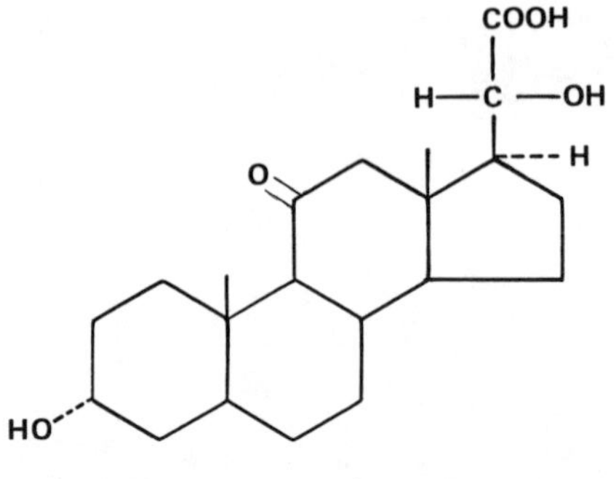

COOH | H—C—OH | ---H ... HO

20 β - Deoxycortolonic acid

their stereochemistry at position 20.[270] The discovery of these acids, although unexpected, is consistent with known chemical transformations. Mattox[271] has shown that in acid solution the side chain of 17-hydroxy corticosteroids is transformed by dehydration and rearrangement to a 17-deoxy-21-dehydro side chain. The metabolic pathway leading to a similar end product is shown in Figure 23. The figure shows two alternative sequences. Although several of the steps correspond with metabolic transformations which we have already demonstrated, the formation and metabolism of intermediate III has not yet been shown. The alternative possibility that the 17-deoxy steroid acids arose as artifacts of isolation or of contamination appears to be extremely unlikely. Although the existence of the new pathway is clearly established, its physiological significance remains to be determined.

iii. Analytical Procedures

Having established the existence of the cortoic acids as quantitatively important urinary products of cortisol, the need for a convenient assay procedure for these compounds became essential. Assay methods which utilize chromatographic separation and microanalytical detection are, for the most part, universally applicable. Alternatively, radioimmunoassay procedures provide the advantage of high sensitivity, al-

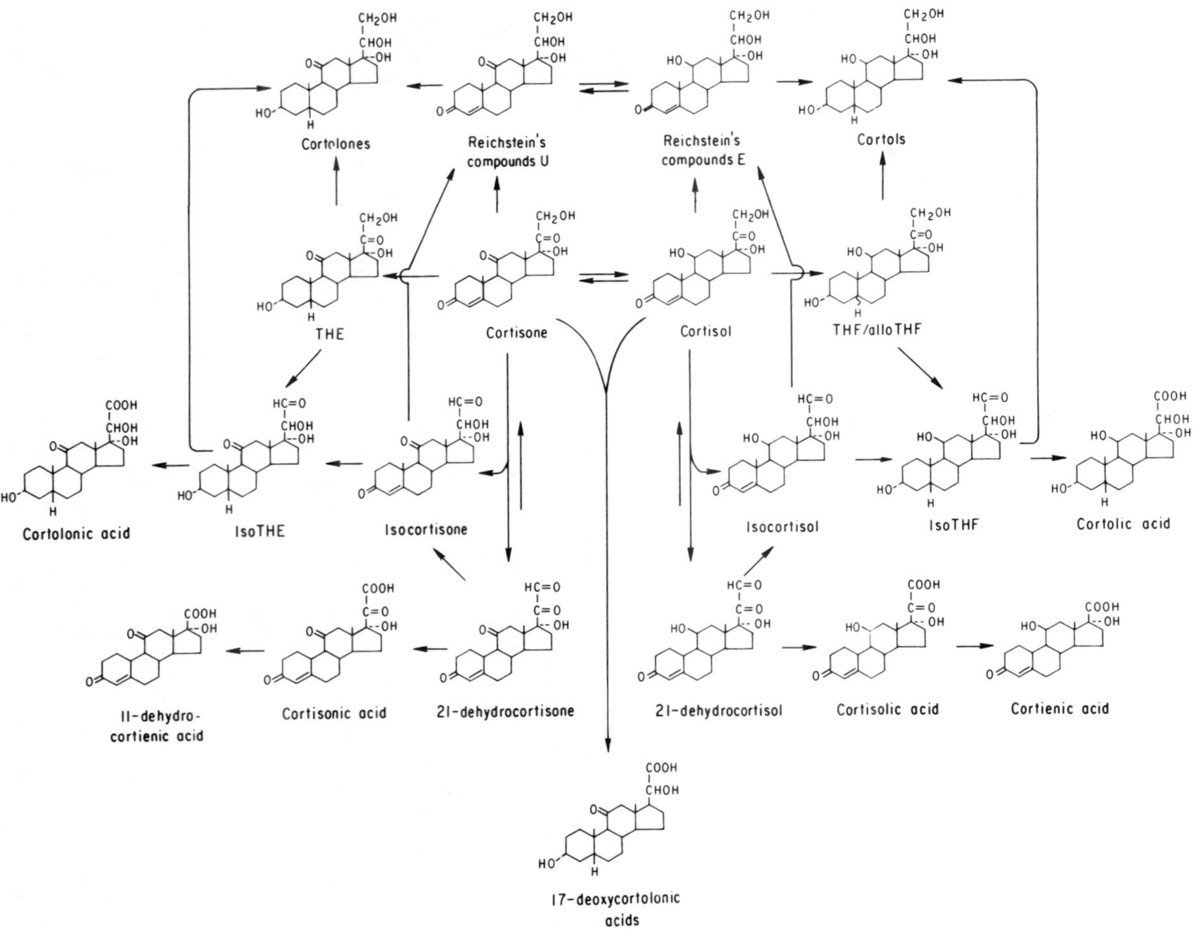

FIGURE 22. Pathways of cortisol metabolism.

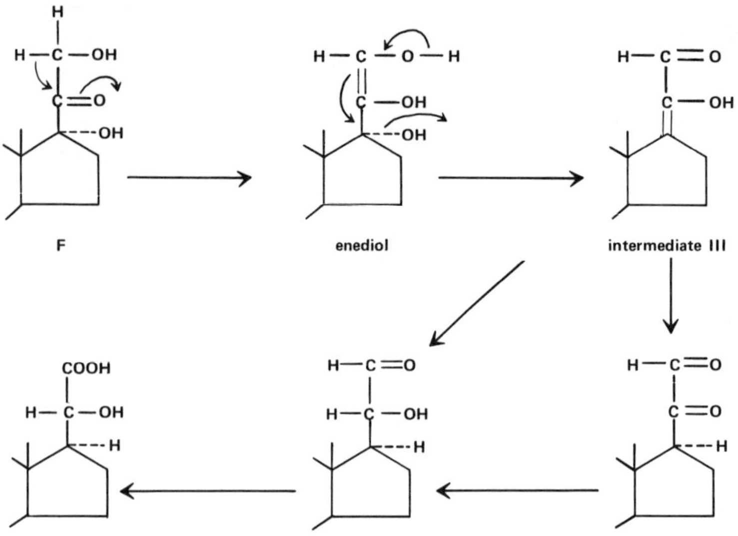

FIGURE 23. Postulated pathway of 17-deoxycortoic acid biosynthesis.

though they do not provide the versatility for the analysis of large numbers of different acids. Because of the respective advantages of each approach, we have been exploring both.

(a) Chromatographic Procedures

Attempts to develop a fluorometric assay based on our earlier observations on the formation of a fluorescent derivative of etienic acid[154] were unsuccessful. Gas-liquid chromatography, though potentially useful, proved to be cumbersome and time-consuming. Derivatization was complex and tedious. Retention times on commonly available supports were excessively long. High-pressure liquid chromatography (HPLC) is a convenient, simple, quantitative, nondestructive procedure that appeared to be very well suited to the analysis of the cortoic acids. Its utility is restricted by the optical transparency of the urinary metabolites at 254 nm, the wavelength commonly used in fixed wavelength spectrophotometric flow detectors. Derivatization with suitable chromophores was therefore necessary. Durst et al.[272] have developed a technique for visualizing fatty acids in effluents from HPLC columns by coupling the acids with α,*p*-dibromoacetophoenone using crown ethers as catalyst. The resulting *p*-bromophenacyl esters absorb at 254 nm. We have adapted their method to the analysis of cortoic acids.[273] The phenacyl esters were readily separated in HPLC (Figure 24). The method is quantitative and is capable of measuring 0.5 pmol of acid.

(b) Radioimmunoassay

Both epimeric cortolonic acids have been coupled to bovine serum albumin and antibodies to the complexes raised in rabbits. Sera usable at a dilution of 1:1000 have been obtained. Either (1,2^3H) cortolonic acid or the (I^{125}) tyrosyl derivatives of the acids have been used as the labeled ligand. The method is sensitive to 50 pg of acid. Cross reactivity was low and specificity satisfactory.

(c) Other Methods

An alternative approach modeled after the Norymberski procedure has been proposed by Senciall et al.[274] The acidic fraction is reduced with sodium borohydride and cleaved with periodic acid to give 17-ketosteroid, which is then measured by the Zimmerman reaction. The method is specific for 17-hydroxy acids.

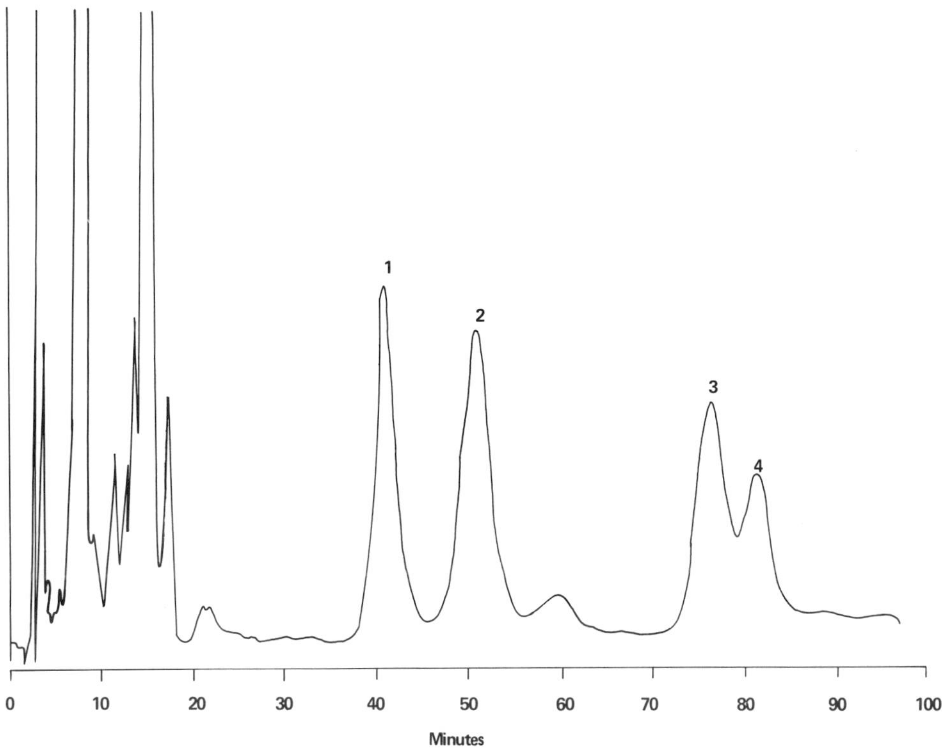

FIGURE 24. HPLC separation of cortoic acids. Separation of *p*-bromophenacyl esters was obtained on Zorbax ODS columns (4.6 × 250 mm) with methanol: water; 65:35 as mobile phase. Flow rate 1.5 cc/min; 4500 psi; 40°C. The steroids are 1 cortolic acid, 2 cortolonic acid, 3 β-cortolic acid, and 4 β-cortolonic acid.

TABLE 13

Excretion of Cortoic Acids by Patients with Different Disease States

Class of patient	Cortoic Acids % of Dose
Normal	6.2 ± 1.1 (22)[a]
Anorexia nervosa	4.8 ± 0.4 (3)
Coronary disease	8.3 ± 1.5 (4)
Acute intermittent porphyria	31 ± 2.5 (6)
Breast Cancer	
Low acid formers	5.6 ± 0.9 (23)
Intermed. acid formers	9.6 ± 2.2 (35)
High acid formers	16.3 ± 2.7 (55)
Prostate cancer	
Low acid formers	3.7 ± 1.1 (10)
Intermediate acid formers	7.8 ± 1.7 (15)
High acid formers	16.0 ± 3.1 (31)

[a] Number of patients in parentheses.

(d) Urinary Levels of Acids in Humans

The determination of the extent of formation of steroid acids is critical to any assessment of their physiological role in disease. We have measured total radioactive acid formation in a variety of patients who received (4-^{14}C)-cortisol (Table 13). Marked

TABLE 14

Comparison of Cortoic Acid Excretion from Patients with Different Diseases

	17-Deoxy cortolonic acids	Cortolonic acids		Cortolic acids	
		α	β	α	β
Cancer of the Prostate	1.6	2.4	3.6	0.3	1.5
Coronary	1.2	2.4	3.6	0.3	1.3
Normal	0	2.8	4.1	0.2	1.2
Anorexia	1.0	0.7	0.4	0.5	1.8

difference in the relative yield of acids was observed in a number of clinical conditions. Patients with anorexia nervosa excreted 4.8% of the dose as acids, compared with normal subjects whose acid excretion averaged 6%. The most striking increase was observed with a patient suffering from acute intermittent porphyria who excreted fully one third of the administered tracer as acids. Of particular interest was the pattern of excretion of cortoic acids by patients with prostatic cancer and breast cancer. For each category the levels fell into three distinct patterns: normal, intermediate, and high. This variability may be a function of a number of extrinsic factors unrelated to the disease process, as well as being a reflection of the disease itself. It is possible that the apparently normal levels of acid in some of the cancer patients may obscure an abnormal distribution of acids in these patients. As Table 14 shows, this does not appear to be the case. The profile of individual steroid acids in the case of a patient with prostatic cancer whose total excretion was within the normal range appeared to be within normal limits.

iv. Formation of Acids from Deoxycorticosterone (DOC)

The excretion of acidic metabolites of DOC into the urine in humans and several other species has been reported by Senciall.[210] Although the conventionally labeled steroid used by these workers provides valuable information for comparative interspecies excretion of acidic DOC metabolites, more information can be obtained by the use of (21^3H)-DOC. In humans, DOC, like cortisol, is converted to acidic metabolites, albeit in lower yield.[275] The release of tritium into the body water was greater than could be accounted for by acid formation. These results are expected, in view of our in vitro observations with model systems, where the 21-methylene tritium of DOC was enzymatically exchanged with the protium of water. No 20-tritiated acid metabolites could be isolated. This suggests the possibility that the metabolic transformations of 17-deoxy corticosteroids may differ in some respects from those of the 17-hydroxy corticosteroids. The absence of a 17-hydroxy group appears to decrease the magnitude of the side chain oxidation and may possibly lead to the preferential formation of 20-oxo acids.

REFERENCES

1. **Bloch, K.**, Biogenesis and transformation of squalene, in *Ciba Symposium on Biosynthesis of Terpenes and Steroids*, Little, Brown, Boston, 1959, 70.
2. **Miller, W. L. and Gaylor, J. L.**, Investigation of the component reactions of oxidative sterol demethylation oxidation of a 4α-methyl sterol to a 4α-carboxylic acid during cholesterol biosynthesis, *J. Biol. Chem.*, 245, 5369, 1970.
3. **Olson, J. A., Jr., Lindberg, M., and Bloch, K.**, On the demethylation of lanosterol to cholesterol, *J. Biol. Chem.*, 226, 941, 1957.
4. **Bloch, K.**, The biological synthesis of cholesterol, *Science*, 150, 19, 1965.
5. **Mulheirn, L. J. and Ramm, P. J.**, The biosynthesis of sterols, *Chem. Soc. Rev.*, 1, 259, 1972.
6. **Lindberg, M., Gautschi, F., and Bloch, K.**, Ketonic intermediates in the demethylation of lanosterol, *J. Biol. Chem.*, 238, 1661, 1963.
7. **Swendell, A. C. and Gaylor, J. L.**, Investigation of the component reactions of oxidative sterol demethylation. Formation and metabolism of 3-ketosteroid intermediates, *J. Biol Chem.*, 243, 5546, 1968.
8. **Wilton, D. C. and Akhtar, M.**, The mechanism of C-4 demethylation during cholesterol biosynthesis, *Biochem. J.*, 149, 233, 1975.
9. **Gibbons, G. F. and Mitropoulos, K. A.**, Effect of trans-1,4 bis (2-Chlorobenzylaminomethyl) cyclohexane dihydrochloride and carbon monoxide on hepatic cholesterol biosynthesis from 4,4-dimethyl sterols *in vitro*, *Biochim. Biophys. Acta*, 380, 270, 1975.
10. **Gaylor, J. L., Meyake, Y., and Yamano, T.**, Stoichiometry of 4-methyl sterol oxidase of rat liver microsomes, *J. Biol. Chem.*, 250, 7159, 1975.
11. **Bechtold, M. M., Delevich, C. V., Comai, K., and Gaylor, J. L.**, Investigation of the component reactions of oxidative sterol demethylation. Role of an endogenous microsomal source of reducing equivalents, *J. Biol. Chem.*, 247, 7650, 1972.
12. **Bloxham, D. P. and Akhtar, M.**, Studies on the control of cholesterol biosynthesis: the adenosine 3′:5′-cyclic monophosphate-dependent accumulation of a steroid carboxylic acid, *Biochem. J.*, 123, 275, 1971.
13. **Bloxham, D. P., Wilton, D. C., and Akhtar, M.**, Studies on the mechanism and regulation of C-4 demethylation in cholesterol biosynthesis. The role of adenosine 3′:5′ cyclic monophosphate, *Biochem. J.*, 125, 625, 1971.
14. **Miller, W. L. and Taylor, J. L.**, Investigation of the component reactions of oxidative sterol demethylation: oxidation of a 4,4 dimethyl sterol to a 4β-methyl-4α-carboxylic acid during cholesterol biosynthesis, *J. Biol. Chem.*, 245, 5375, 1970.
15. **Brady, D. R.**, Sterol aldehyde intermediates in cholesterol biosynthesis. Metabolism of 4-hydroxymethylene-5α-cholest-7-en-3-one. Enzymic formation of esters of methylsterol precursors of cholesterol, an enzymic process that competes with cholesterol biosynthesis, *Diss. Abstr. Int. B*, 31, 7078, 1971.
16. **Nelson, J. A., Kahn, S., Spencer, T. A., Sharpless, K. B., and Clayton, R. B.**, Some aspects of substrate specificity in biological demethylation at C_4 of steroids, *Bioorg. Chem.*, 4, 363, 1975.
17. **Miyake, Y. and Gaylor, J. L.**, Electron transfer from α-reduced nicotinamide adenine dinucleotide to flavoprotein, cytochromes, and mixed function oxidases of rat liver microsomes, *J. Biol. Chem.*, 248, 7345, 1973.
18. **Hornby, G. M. and Boyd, G. S.**, A carboxylic acid intermediate of lanosterol demethylation, *Biochem. Biophys. Res. Commun.*, 40, 1452, 1970.
19. **Gaylor, J. L. and Delwiche, C. V.**, Investigation of the multienzymic system of microsomal cholesterol biosynthesis, *Ann. N.Y. Acad. Sci.*, 212, 122, 1973.
20. **Nes, W. R.**, Regulation of the sequencing in sterol biosynthesis, *Lipids*, 6, 219, 1971.
21. **Gibbons, G. F. and Mitropoulis, K. O.**, The effect of CO on the nature of the accumulated 4,4-dimethyl sterol precursors of cholesterol during its biosynthesis from (2-^{14}C) mevalonic acid *in vitro*, *Biochem. J.*, 132, 439, 1973.
22. **Fiecchi, A., Galli Kienle, M., Scala, A., Galli, G., Grossi Paoletti, E., Cattebini, F., and Paoletti, R.**, Hydrogen exchange and double bond formation in cholesterol biosynthesis, *Proc. R. Soc. London Ser. B*, 180, 125, 1972.
23. **Akhtar, M., Wilton, D. C., Watkinson, I. A., and Rahimtula, A. D.**, Substrate activation in pyridine nucleotide-linked reactions: illustrations from the steroid field, *Proc. R. Soc. London, Ser. B*, 180, 167, 1972.
24. **Alexander, K. T. W., Akhtar, M., Boar, R. B., McGhie, J. F., and Barton, D. H. R.**, Removal of the 32-c-atom as formic acid in cholesterol biosynthesis, *J. Chem. Soc. Chem. Commun.*, 383, 1972.
25. **Gibbons, G. F.**, The metabolic sequence by which some 4,4-dimethyl sterols are converted into cholesterol, *Biochem. J.*, 144, 59, 1974.

26. **Gibbons, G. F. and Ramansanda, K.**, Synthesis and configuration at C-15 of the epimeric 5α-androst-8-ene-3β,5α-diols, *J. Chem. Soc. Chem. Commun.*, 213, 1975.
27. **Kariyone, T. and Kurono, G.**, Constitutents of *formes officinalis fris.*, *J. Pharm. Soc. Jpn.*, 60, 318, 1940.
28. **Simonsen, J. and Ross, W. Z. J.**, *The Terpenes,* Vol 5, Cambridge University Press, Cambridge 1957.
29. **Gascorgne, R. M., Holker, J. S. E., Ralph, R. J., and Robertson, A.**, Fungi (XVI) eburicoic acid, *Nature (London),* 166, 652, 1952.
30. **Fried, J., Grabowich, P., Sabo, E. F., and Cohen, A. I.**, The structure of sulphurenic acid: a new triterpenoid from polyporus sulphurens, *Tetrahedron,* 20, 2297, 1964.
31. **Cross, L. C., Eliot, C. G., Herbron, I. M., and Jones, E. R. H.**, Constituents of the higher fungi, Part 1. The triterpene acids of *polyporus betulanius fr.*, *J. Chem. Soc.*, p. 632, 1940.
32. **Barton, D. H. R. and Jones, E. R. H.**, Optical rotary power and structure in triterpenoid compounds application of the method of molecular rotation differences, *J. Chem. Soc.*, p. 659, 1944.
33. **Halsall, T. G., Jones, E. R. H., and Lemin, A. J.**, The chemistry of triterpenes. XV. The environment of the unreactive double bond of polyporenic acid A, *J. Chem. Soc.*, p. 468, 1953.
34. **Halsall, T. G., Hodges, R., and Jones, E. R. H.**, The chemistry of the triterpenes and related compounds. XIX. Further evidence concerning the structure of polyporenic acid A, *J. Chem. Soc.*, p. 3019, 1953.
35. **Guider, J. M., Halsall, T. G., Hodges, R., and Jones, E. R. H.**, The chemistry of triterpenes and related compounds. XXVI. The nature of polyporenic acid B, *J. Chem. Soc.*, p. 3234, 1954.
36. **Frerejaque, M.**, Note sur I'acide ungulinique acid cristallise isolé de polyporus (ungulina) *betulinus fr.*, *Rev. Mycol.*, 3, 95, 1938.
37. **Marcus, S.**, Antibacterial activity of the triterpenoid acid (polyporenic acid C) and of ungulinic acid, metabolic products of *polyporus benzoinus (Wahl) fr.*, *Biochem. J.*, 50, 516, 1952.
38. **Cross, L. C. and Jones, E. R. H.**, Constitutents of the higher fungi. II. The unsaturated system of polyporenic acid A, *J. Chem. Soc.*, p. 1491, 1940.
39. **Jones, E. R. H. and Woods, G. F.**, The chemistry of triterpenes. XIV. Further evidence concerning the unsaturated centres of polyporenic acid A, *J. Chem. Soc.*, p. 464, 1953.
40. **Cross, L. C., Eliot, C. G., Herbron, I. M., and Jones, E. R. H.**, Constituents of higher fungi. I. The triterpene acids of polyporus betulinus fr., *J. Chem. Soc.*, p. 632, 1940.
41. **Guider, J. M., Halsall, T. G., and Jones, E. R. H.**, The chemistry of triterpenes and related compounds. XXVII. Pinicolic acid A, *J. Chem. Soc.*, p. 4471, 1954.
42. **Beereboom, J. J., Fazakereley, H., and Halsall, T. G.**, The chemistry of triterpenes and related compounds. XXXI. The isolation of further triterpene acids from polyporus pinicola fr., *J. Chem. Soc.*, p. 3437, 1957.
43. **Ruzicka, L. and Häusermann, H.**, Zur Kenntnis der Triterpene uber die β-elemolsäure, *Helv. Chim. Acta,* 25, 439, 1942.
44. **Ruzicka, L., Rey, E., and Spillman, M.**, Zur Kenntnis der triterpene uber die α elemolsäure, *Helv. Chim. Acta,* 25, 1375, 1942.
45. **Ruzicka, L., Ray, E., and Spillman, M.**, Zur Kenntnis der triterpene uber die β-elemonsäure, *Helv. Chim. Acta,* 25, 1403, 1942.
46. **Belham, P. and Kou, G. A. R.**, Sapogenins, XVI. The acids of elemi resin, *J. Chem. Soc.*, p. 544, 1942.
47. **Barton, D. H. R. and Seoane, E.**, Triterpenoids. XXII. The constitution and stereochemistry of masticadienonic acid, *J. Chem. Soc.* p. 4150, 1956.
48. **Seoane, E.**, Further crystalline constituents of gum mastic, *J. Chem. Soc.*, p. 4158, 1956.
49. **Fried, J., Krakower, G. W., Rosenthal, D., and Basch, H.**, Structure-activity relations in the field of antibacterial steroid acids, *J. Med. Chem.*, 8, 279, 1965.
50. **Godtfredsen, W. O. and Vangedal, S.**, Structure of fusidic acid, *Tetrahedron,* 18, 1024, 1962.
51. **Arigoni, D., von Daehne, W., Godtfredsen, W. O., Melera, A., and Vangedal, S.**, The stereochemistry of fusidic acid, *Experientia,* 20, 344, 1964.
52. **Allinger, N. L. and Friedberg, L. A.**, The synthesis of highly strained medium rings, *J. Org. Chem.*, 26, 4552, 1961.
53. **Okada, S., Iwasaki, S., Tsuda, K., Sano, Y., Hato, K., Udagawa, S., Nakayama, Y., and Yamaguchi, H.**, Structure of helvolic acid, *Chem. Pharm. Bull.*, 12, 121, 1964.
54. **Barton, D. H. R., Pardhaus, K., Sternhill, S., and Templeton, J. F.**, Triterpenoids. XXV. The constitutions of limonic and related bitter principles, *J. Chem. Soc.*, p. 255, 1961.
55. **Baird, B. M., Halsall, T. G., Jones, E. R. H., and Lowe, G.**, Cephalosporin, P., *Proc. Chem. Soc., London,* p. 257, 1961.
56. **Haslewood, G. A. D.**, *Bile Salts,* Methuen, London, 1967.
57. **Enomoto, S.**, Sterobile acids and bile steroids. XLV. The metabolism of 3α,7α,12α-trihydroxy coprostane in toad liver homogenate, *J. Biochem.*, 52, 1, 1962.

58. **Vanderah, D. J. and Djerassi, C.**, Novel marine sterols with modified bile acid side chains from the sea pen *Ptilosarius gurneyi, Tetrahedron Lett.*, 8, 683, 1977.
59. **Mandava, N., Anderson, J. D., Dutky, S. R., and Thompson, M. J.**, Novel occurence of 5β-cholanic acid in plants; isolation from jequirity bean seeds *(Abrus precatvius L.), Steroids,* 23, 357, 1974.
60. **Ziegler, R. and Tamm, C.**, Isolation and structure of eucosterol and 16β-hydroxy eucosterol, two novel spirocyclic nortriterpens, and of a new 24-nor-5α-chola-8,16-diene-23 oic acid from bulbs of several eucomis species, *Helv. Chim. Acta,* 59, 1997, 1976.
61. **Siperstein, M. D. and Murray, A. W.**, Cholesterol metabolism in man, *J. Clin. Invest.*, 34, 1449, 1955.
62. **van Itallie, T. B. and Hashima, S. A.**, Clinical and experimental aspects of bile acids in man. Hydroxylation of the C_{27} side chain, *Med. Clin. North Am.*, 47, 629, 1963.
63. **Bergstrom, S.**, Metabolism of bile acids, *Fed. Am. Soc. Exp. Biol.*, 20, 121, 1961.
64. **Bjorkhem, I., Gustafsson, J., Johansson, G., and Persson, B.**, Biosynthesis of bile acids in man. Hydroxylation of the C_{27} side chain, *J. Clin. Invest.*, 55, 478, 1955.
65. **Taniguchi, S., Hoshita, N., and Okuda, K.**, Enzymatic characteristics of CO sensitive 26 hydroxylase system for 5β-cholestane-3α,7α,12α-triol in rat liver mitochondria. Intramitochondrial localization, *Eur. J. Biochem.*, 40, 607, 1973.
66. **Beisius, O.**, On the stereochemistry of 26-hydroxylation of cholesterol. Bile acids and steroids, *Acta Chem. Scand.*, 19, 325, 1965.
67. **Fredrickson, D. S.**, The conversion of cholesterol-4-^{14}C to acids and other products by liver mitochondria, *J. Biol. Chem.*, 222, 109, 1956.
68. **Cronholm, T. and Johanson, G.**, Oxidation of 5β-cholestane-3α,7α,12α-triol by rat liver microsomes, *Eur. J. Biochem.*, 16, 373, 1970.
69. **Mendelsohn, D. and Mendelsohn, L.**, The *in vitro* catabolism of cholesterol. A comparison of the formation of 26-hydroxycholesterol and chenodeoxycholic acid from cholesterol in rat liver, *Biochemistry,* 7, 4167, 1968.
70. **Mitropoulos, K. A., Avery, M. D., Myant, N. B., and Gibbons, G. F.**, The formation of cholest-5-ene-3β,26 diol as an intermediate in the conversion of cholesterol into bile acids by liver microsomes, *Biochem. J.*, 130, 363, 1972.
71. **Schwartz, C. C., Cohen, B. I., Vlahcerci, Z. R., Gregory, D. H., Halloran, L. G., Kuramoto, J., Mosbach, E. H., and Swell, L.**, Quantitative aspects of the conversion of 5βcholestane intermediates to bile acids in man, *J. Biol. Chem.*, 251, 6308, 1976.
72. **Bjorkhem, I., Danielsson, H., and Kovall, K.**, Side chain hydroxylation in biosynthesis of cholic acid. 25 and 26 hydroxylation of 5β-cholestane 3α,7α,12α-triol by reconstituted systems from rat liver microsomes, *J. Biol. Chem.*, 251, 3495, 1976.
73. **Salen, G., Shefer, S., Seloguche, T., and Mosbach, E. H.**, Bile alcohol metabolism in man. Conversion of 5β-cholestane-3α,7α,12α,25, tetrol to cholic acid, *J. Clin. Invest.*, 56, 226, 1975.
74. **Danielsson, H. and Sjovall, J.**, Bile acid metabolism, *Annu. Rev. Biochem.*, 44, 233, 1975.
75. **Danielsson, H.**, Mechanisms of bile acid biosynthesis in *The bile acids,* Vol. 2, 1, Nair, P. P. and Krichevsky, D., Eds., Plenum Press, New York, 1973.
76. **Yamada, T.**, Sterobile acids and bile alcohols. Formation of cholic acid from 5β-cholestane-3α,7α,12α,24,26 pentol and 5β-cyprinol in the bile fistula rat, *Hiroshima J. Med. Sci.*, 15, 375, 1966.
77. **Bjorkhem, F., Gustafsson, J., Johansson, G., and Persson, B.**, Hydroxylation of the C_{27} steroid side chain, *J. Clin. Invest.*, 55, 478, 1975.
78. **Dielehy, J. M. and Wilson, J. D.**, Regulation of cholesterol metabolism, *N. Eng. J. Med.*, 282, 1241, 1970.
79. **Bjorkhem, I., Jörnvall, H., and Akeson, A.**, Oxidation of hydroxylated fatty acids and steroids by SS-isoenzyme of liver alcohol dehydrogenase, *Biochem. Biophys. Res. Commun.*, 57, 870, 1974.
80. **Okuda, K. and Takigawa, H.**, Rat liver 5β-cholestane-3α,7α,12α,26-tetrol dehydrogenase as a liver alcohol dehydrogenase, *Biochim. Biophys. Acta,* 220, 141, 1970.
81. **Okuda, K., Higuchi, E., and Fukuba, R.**, Horse liver 3α,7α,12α-trihydroxy-5β-cholestan-26-al dehydrogenase as a liver aldehyde dehydrogenase, *Biochim. Biophys. Acta,* 793, 15, 1973.
82. **Fukuba, R.**, Stereochemistry of hydrogen transfer between pyridine nucleotides and some intermediates of cholesterol catabolism catalysed by liver alcohol and aldehyde dehydrogenase, *Biochim. Biophys. Acta,* 341, 48, 1974.
83. **Masui, T., Herman, R., and Staple, E.**, The oxidation of 5β-cholestane-3α,7α,12α,26-tetrol to 5β-cholestane-3α,7α,12α,triol-26-oic acid via 5β-cholestane-3α,7α,12α-triol-26-al by rat liver, *Biochim. Biophys. Acta,* 117, 266, 1966.
84. **Okuda, K. and Danielsson, H.**, Synthesis and metabolism of 5β-cholestane-3α,7α,12α-triol-26-al to bile acids and steroids, *Acta Chem. Scand.*, 19, 2160, 1965.
85. **Suld, H. M., Staple, E., and Gurin, S.**, Mechanism of formation of bile acids from cholesterol: oxidation of 5β-cholestane-3α,7α,12α-triol and formation of propionic acid from the side chain by rat liver mitochondria, *J. Biol. Chem.*, 237, 338, 1962.

86. **Briggs, T., Whitehouse, M. W., and Staple, E.,** Metabolism of trihydroxy coprostanic acid: Formation from cholesterol in the alligator and conversion to cholic acid and carbon dioxide *in vitro* by rat liver mitochondria, *J. Biol. Chem.,* 236, 688, 1961.
87. **Mosbach, E. H.,** Hepatic synthesis of bile acids. Biochemical steps and mechanisms of rate control, *Arch. Intern. Med.,* 130, 478, 1972.
88. **Masui, T.** and Staple, E., The formation of cholic acid by $3\alpha,7\alpha,12\alpha$,24-tetrahydroxy coprostanic acid by rat liver, *Biochim. Biophys. Acta,* 104, 305, 1965.
89. **Inai, Y., Tahaka, Y., Betsuki, S., and Kazuno, T.,** Sterobile acids and bile sterols. Synthesis of $3\alpha,7\alpha,12\alpha$,24-tetrahydroxycoprostanic acid, *J. Biochem. (Tokyo),* 56, 591, 1964.
90. **Yashima, H.,** Sterobile acids and bile sterols 98. The partial synthesis of norcholic aldehyde and bisnorcholic aldehyde, *Hiroshima J. Med. Sci.,* 16, 21, 1967.
91. **Yashima, H.,** Sterobile acids and bile sterols. Synthesis and metabolism of cholic aldehyde, *J. Biochem. (Tokyo),* 54, 47, 1963.
92. **Talalay, P.,** Enzymatic mechanisms in steroid metabolism, *Physiol. Rev.,* 37, 362, 1957.
93. **Turfitt, G. E.,** Microbiological agencies in the degradation of steroids. I. Cholesterol decomposition organisms of soil, *J. Bacteriol.,* 47, 487, 1944.
94. **Schatz, A., Savard, K., and Pintner, I. J.,** Ability of soil microorganisms to decompose steroids, *J. Bacteriol.,* 58, 117, 1949.
95. **Mills, J. S. and Werner, A. E. A.,** The chemistry of dammar resin, *J. Chem. Soc.,* p. 3132, 1955.
96. **Arigoni, D., Barton, D. H. R., Bernassioni, R., Djerassi, C., Mills, J. S., and Wolff, R. E.,** The constitutions of dammaretolic and nyctanthic acid, *J. Chem. Soc.,* p. 1900, 1960.
97. **Mangoni, L. and Belardini, M.,** Constituents of Grindelia robusta. II. 6-Oxygrindilic acid III $7\alpha,8\alpha$ oxidohydrogrindilic acid, *Tetrahedron Lett.,* p. 921, 1963.
98. **Hayakawa, S. and Fujiwara, T.,** Microbiological degradation of bile acid, *Biochem. J.,* 162, 387, 1977.
99. **Hayakawa, S. and Hashimoto, S.,** (+)-(5R)-Methyl-4-oxo-octane-1,8-dione acid, a microbiological degradation product from rings C and D of cholic acid, *Biochem. J.,* 12, 127, 1969.
100. **Fried, J., Thoma, R. W., and Klingsberg, A.,** Oxidation of steroids by microorganisms. II. Side chain degradation of ring D cleavage and dehydrogenation in ring A, *J. Am. Chem. Soc.,* 75, 5764, 1953.
101. **Peterson, A. K., Eppstein, S. H., Meister, P. D., Murray, H. C., Leigh, H. M., Weinstraub, M., and Runeike, L. M.,** Microbiological transformations of steroids. VX. Degradation of C_{21} steroids to C_{19} ketones and to testololuctone. *J. Am. Chem. Soc.,* 75, 5768, 1953.
102. **Tayib, Q. E., Knight, S. G., and Sih, C. J.,** Substrate specificity in steroid side-chain degradation by microorganisms, *Biochim. Biophys. Acta,* 93, 411, 1964.
103. **Turfitt, G. E.,** Microbiological Degradation of Steroids. IV. Fission of the steroid molecule, *Biochem. J.,* 42, 376, 1948.
104. **Stadtman, T. C., Cherkes, A., and Anfinsen, C. B.,** Studies on the microbiological degradation of cholesterol, *J. Biol. Chem.,* 206, 511, 1954.
105. **Santer, M. and Ajl, S. J.,** Steroid metabolism by a species of Pseudomonas. II. Direct evidence for the breakdown of testosterone, *J. Biol. Chem.,* 199, 85, 1952.
106. **Coombe, R. G., Tsong, Y. Y., Hamilton, P. B., and Sih, C. J.,** Mechanisms of steroid oxidation by microorganisms X. Oxidative cleavage of estrone, *J. Biol. Chem.,* 241, 1587, 1968.
107. **Dodsen, R. M. and Muir, R. D.,** Microbiological transformations. IV. The microbiological aromatization of steroids, *J. Am. Chem. Soc.,* 83, 4627, 1961.
108. **Shaw, D. A., Bockenhagen, L. F., and Talalay, P.,** Enzymatic oxidation of steroids by cell-free extracts of Pseudomonas testosteroni: isolation of cleavage products of ring A, *Proc. Natl. Acad. Sci., U.S.A.,* 54, 837, 1964.
109. **Schubert, K., Bohme K-H., Ritter, F., and Horhold, C.,** Mikrobeller Abbau von Progesteron zu α-Ketoglutarsaure und Bernsteinsaure, *Biochim. Biophys. Acta,* 152, 401, 1968.
110. **Sih, C. J.,** Mechanisms of steroid oxidation by microorganisms, *Biochim. Biophys. Acta,* 62, 341, 1962.
111. **Sih, C. J., Lee, S. S. Tsong, Y. Y., and Wang, K. C.,** Mechanisms of steroid oxidation by microorganisms. VIII. 3,4-Dihydroxy-9,10-secoandrosta-1,3,5(10)-triene-9,17-dione, an intermediate in the microbiological degradation of ring A of androst-4-ene-3,17-dione, *J. Biol. Chem.,* 241, 540, 1966.
112. **Gibson, D. T., Wang, K. C., Sih, C. J., and Whitlock, H., Jr.,** Mechanisms of steroid oxidation by microorganisms, IX. On the mechanism of ring A cleavage in the degradation of 9,10-seco steroids by microorganisms, *J. Biol. Chem.,* 241, 551, 1966.
113. **Dagley, S., Chapman, P. J., Gibson, D. T., and Wood, J. M.,** Degradation of the benzene nucleus by bacteria, *Nature (London),* 202, 775, 1964.
114. **Hayaishi, O., Katagiri, M., and Rothberg, S.,** Studies on oxygenases, *J. Biol. Chem.,* 229, 905, 1957.
115. **Tai, H. H. and Sih, C. J.,** 3,4-Dihydroxy-9,10-secoandrosta-1,3,5(10)-triene-9,17-dione 4,5-dioxygenase from Nocardia restrictus. II. Kinetic studies, *J. Biol. Chem.,* 245, 5072, 1970.

116. **Howe, R., Moore, R. H., Rao, B. S., and Gibson, R. I.,** Microbial modification of diosgenin and hecogenin, *J. Chem. Soc.*, p. 1940, 1973.
117. **Laskin, A., Grabowitch, P., Meyers, C. D., and Fried, J.,** Transformation of eburicoic acid. V. Cleavage of ring A by the fungus *Glomerella fusarioides, J. Med. Chem.*, 7, 406, 1964.
118. **Schubert, K., Böhme, R. H., and Horhold, C.,** Bildung einer Ketosaure durch mikrobieller Abbau von Progesteron, *Z. Physiol. Chem.*, 325, 260, 1961.
119. **Schubert, K., Böhme K-H., and Horbold, C.,** Mikrobieller abbau-von desoxy corticosterone zu 7α-methyl-1 glykol-perhydroindanon-5-β-propiansaure-4, *Z. Naturforsch.*, 186, 988, 1963.
120. **Tan, T.-L., Strijewski, A., and Wagner, F.,** Degradation of 5-pregnene 3β,20β diol by a *Nocardia Sp., Arch Microbiol.*, 87, 249, 1972.
121. **Hayakawa, S., Kanenalau, Y., and Fujiwara, T.,** Microbiological degradation of bile acids, *Biochem. J.*, 115, 249, 1964.
122. **Peterson, R. E.,** Metabolism of adrenal cortical steroids, in *The Human Adrenal Cortex,* Christy, N. P., Ed., Harper and Row, New York, 1971, 109.
123. **Brooks, R. V.,** The metabolism of cortisol in Cushing's disease, in *Structure and Metabolism of Corticosteroids,* Pasqualini, J. and Jayle, M. F., Eds., Academic Press, New York, 1964, 119.
124. **Zumoff, B., Bradlow, H. L., Gallagher, T. F., and Hellman, L.,** Cortisol metabolism in cirrhosis, *J. Clin. Invest.*, 46, 1735, 1967.
125. **Migeon C. J., Sandberg, A. A., Decker, H. A., Smith, D. F., Paul, A. C., and Samuels, L. T.,** Metabolism of 4-^{14}C-cortisol in man: body distribution and rates of conjuation, *J. Clin. Endocrin.*, 16, 1137, 1956.
126. **Peterson, R. E., Wyngaarden, J. B., Guerra, S. L., Brodie, B. B., and Bunim, J. J.,** The physiological disposition and metabolic fate of hydrocortisone in man, *J. Clin. Invest.*, 34, 1779, 1955.
127. **Southcott, C. M., Gandossi, S. K., Barker, A. D., Bandy, H. E., McIntosh, H., and Darrach, M.,** Human adrenal steroids. I. The effect of corticotropin on components of the free and conjugated plasma C_{21} adrenal steroid fraction, *Can. J. Biochem. Physiol.* 34, 146, 1956.
128. **Pepe, G. J. and Townsley, J. D.,** Cortisol metabolism in female baboons, *Endocrinology,* 95, 1658, 1974.
129. **Lowy, J., Albepart, T. and Pasqualini, J. R.,** ^{3}H-Corticosterone metabolism in the rat, *Acta. Endocrinol. (Copenhagen),* 61, 483, 1969.
130. **Gatehouse, P. W., Roy, A. B., Dodgson, K. S., Powell, G. M., Lloyd, A. G., and Olausen, A. H.,** The metabolism of sodium-cortisone 21-(^{35}S-) sulfate in the rat, *Biochem. J.*, 127, 661, 1972.
131. **Taylor, W.,** Steroid metabolism in the cat. Biliary and urinary excretion of metabolites of (4-^{14}C) cortisone, *Biochem. J.*, 113, 259, 1969.
133. **Gray, C. H. and Shaw, D. A.,** The metabolism of (4-^{14}C) cortisol in patients with sarcoidosis, *J. Endocrinol.*, 33, 57, 1965.
134. **Gray, C. H. and Shaw, D. A.,** The metabolism of (4-^{14}C) cortisol in patients with collagen disease, *J. Endorinol.*, 33, 33, 1965.
135. **Mason, H. L.,** Studies in the chemistry of the adrenal cortex: conversion of compound E to the $C_{21}O_4$ series and to androsterone by the action of calcium hydroxide, *Proc. Staff meet. Mayo Clin.*, 13, 235, 1938.
136. **Mason, H. L.,** Chemical studies of the suprarenal cortex. V. Conversion of compound E to the series which contains four atoms of oxygen and to adrenosterone by the action of calcium hydroxide, *J. Biol. Chem.*, 124, 475, 1938.
137. **Wendler, N. L. and Graber, R. P.,** Alkaline degradation of the cortisol side chain, *Chem. Ind. (London),* p. 549, 1956.
138. **Monder, C.,** Stability of corticosteroids in aqueous solutions, *Endocrinology,* 82, 318, 1968.
139. **Lewbart, M. L. and Mattox, V. R.,** Destruction of cortisone and related steroids by traces of copper during purification procedures, *Nature (London),* 183, 820, 1959.
140. **Oesterling, T. D. and Guttman, D. I.,** Factors influencing stability of prednisolone in aqueous solution, *J. Pharm. Sci.*, 53, 1189, 1964.
141. **Guttman, D. E. and Meister, P. D.,** Kinetics of the base-catalysed degradation of prednisolone, *J. Am. Pharm. Assoc. Sci. Ed.*, 47, 773, 1958.
142. **Olson, M. C.,** Analysis of adrenocortical steroids in pharmaceutical preparations by high pressure liquid-liquid chromatography, *J. Pharm. Sci.*, 62, 2001, 1973.
143. **Herzeny, P. T. and Ehrenstein, M.,** Investigations of steroids. XIV. Studies on the series of the adrenal cortical hormones; dehydration of 6β,21-diacetoxy-allopregnan-5-ol-3,20-dione, *J. Org. Chem.*, 16, 1050, 1951.
144. **Chulski, T. and Forest, A. A.,** Effect of solid buffering agents in aqueous suspensions on prednisolone, *J. Am. Pharm. Assoc. Sci. Ed.*, 47, 553, 1958.
145. **Velluz, L., Petit, A., Pesez, M., and Bettet, R.,** Saponification of the acetate of desoxycorticosterone, *Bull. Soc. Chim. (Fr.),* p. 123, 1947.
146. **Kripalani, K. J. and Sorby, D. L.,** Binding of cortisol and its degradation products by human serum albumin, *J. Pharm. Sci.*, 56, 687, 1967.

147. **Monder, C. and Walker, M. C.**, Identification of 11β,17-dihydroxy-3,20-dioxo-4-pregnen-21-al 1,2-^{3}H,21-dehydrocortisol-1,2-^{3}H and 11β-hydroxy-4-androstene-3,17-dione-1,2-^{3}H as contaminants in preparations of cortisol-1,2-^{3}H, *Steroids*, 15, 1, 1970.
148. **Monder, C.**, Stability of corticosteroids in aqueous solutions, *Endocrinology*, 82, 318, 1968.
149. **Westphal, V., Chader, G. J., and Harding, G. B.**, Decomposition of cortisol-4-^{14}C to 11β-hydroxy-androst-4-ene-3,17-dione-4-^{14}C, *Steroids*, 8, 155, 1967.
150. **Monder, C.**, A new approach to the characterization of steroids containing the 17α-ketol side chain, *Biochem. J.*, 90, 522, 1964.
151. **Kittinger, G. W.**, Quantitative gas chromatography of 17-deoxycorticosteroids and other steroids produced by the rat adrenal gland, *Steroids*, 3, 21, 1964.
152. **Norymburski, J. K.**, Evaluation of gas-liquid chromatography and other techniques for the estimation of corticosteroids, *Clin. Chim. Acta*, 34, 187, 1971.
153. **Sobrinho, L. G.**, The sulfuric acid induced fluorescence of androst-4-ene-11β,17αdihydroxy-3-oxo-17 carboxylic acid, *Steroids*, 7, 289, 1966.
154. **Monder, L. and Kendall, J.**, Sulfuric acid induced fluorescence of corticosteroids. Effect of position substituents on fluorescence, *Anal. Biochem.*, 68, 248, 1975.
155. **Le Gaillard, F., Racadot, A., Racadot-Leroy, N., and Dautrevaux, M.**, Isolement de la transcortine humaine par chromatographie d'affinité, *Biochimie*, 56, 99, 1974.
156. **Nussbaum, A. L., Yuan, E. P., Robinson, C. N., Mitchell, A., Oliveto, E. P., Beaton, J. M., and Barton, D. H. R.**, The photolysis of organic nitrites. VII. Fragmentation of the steroidal side chain, *J. Org. Chem.*, 27, 20, 1962.
157. **Wilds, A. L., Harnik, M., Shimizer, R. Z., and Tyner, D. A.**, Methods for total synthesis of steroids. XVIII. Δ^{14}-16-keto steroid approach to ring D. Introduction of 7-carboxy group. Synthesis of 14α,17β and 14β-17α isomers of rac-estra 5(10)6,8-triene-17 carboxylic acid, *J. Am. Chem. Soc.*, 88, 799, 1966.
158. **Hoehn, W. M. and Mason, H. L.**, Degradation of desoxycholic acid to etiodesoxycholic acid through etiodesoxycholyl methyl ketone, *J. Am. Chem. Soc.*, 60, 1463, 1938.
159. **Marker, R. E. and Wittle, E. L.**, Sterols. LXIII. Etiocholanic acids from the pregnanediols, *J. Am. Chem. Soc.*, 61, 1329, 1939.
160. **King, L. C.**, Preparation of 21-pyridinium-3β-hydroxy-5-pregnen-20-one halides and 3β-hydroxy-5-androstene-17-carboxylic acid, *J. Am. Chem. Soc.*, 66, 1612, 1944.
161. **Grundy, H. M., Simpson, S. A., Tait, J. F., and Woodford, M.**, Isolation of further studies on the properties of a highly active mineralcorticoid from beef adrenal extract, *Acta Endocrinol. (Copenhagen)*, 11, 199, 1932.
162. **Marker, R. E., Rohrman, E., Wittle, E. L., Crooks, Jr., H. M., and Jones, E. M.**, Sterols, XCIV. Persulfate oxidation of allopregnane derivatives. *J. Am. Chem. Soc.*, 62, 650, 1940.
163. **Picha, G. M., Saunders, F. J., and Green, D. M.**, An oxidative metabolite of desoxycorticosterone, *Science*, 115, 704, 1952.
164. **Levy, H. and Maloney, P. J.**, The isolation of 3-keto-4-etienic acid and allopregnan-21-ol-3,20-dione from bovine adrenal perfusions of deoxycorticosterone, *Biochim. Biophys. Acta*, 57, 149, 1962.
165. **Ward, P. J. and Birmingham, M. K.**, Hydroxylation of steroids by the isolated rat adrenal, *Acta Endocrinol. (Copenhagen)*, 39, 110, 1962.
166. **Schneider, J. J.**, In vitro conversion of deoxycorticosterone to some acidic metabolites, in *Hormonal Steroids*, Martini, J., and Pecile, A., Ed., Academic Press, New York, 1964, 127.
167. **Neher, R. and Wettstein, A.**, Biosynthesis of aldosterone. Isolation of C_{20}-steroid-lactones from bovine adrenal, *Helv. Chim. Acta*, 43, 623, 1960.
168. **Neher, R. and Wettstein, A.**, Isolation and identification of new adrenal steroids. Characterization of additional materials contained therein, *Helv. Chim. Acta*, 43, 1171, 1960.
169. **Levy, H., Cargill, D. I., Cha, C. H., Hood, B., and Carlo, J. J.**, The isolation of 18-hydroxy-3-oxo-androst-4-ene-17β-carboxylic acid-20,18-lactone, its 11β-hydroxy analog and other substances from bovine adrenal perfusions of deoxycorticosterone, *Steroids*, 5, 131, 1965.
170. **Levy, H., Cha, C. H., Cargill, I., and Carlo, J. J.**, The inhibition of 11β-hydroxylation of deoxycorticosterone by metopirone in bovine adrenal perfusions, *Steroids*, 5, 147, 1965.
171. **Gontscharow, N. P., Wehrberger, K., and Schubert, K.**, Steroid metabolism in primates. IX. Isolation of 11β,18 dihydroxy-4-androstene-3-one-17β-carboxylic acid lactone (20 → 18) from the adrenal venous blood of *Papio hamadryas*, *J. Steroid Biochem.*, 1, 134, 1970.
172. **Bradlow, H. L. and Monder, C.**, unpublished observations.
173. **Monder, C., Zumoff, B., Bradlow, H. L., and Hellman, L.**, Studies on the biotransformation of cortisol to the cortoic acids in man. Metabolism of 21-dehydrocortisol, *J. Clin. Endocrinol. Metab.*, 40, 86, 1975.
174. **Benes, P., Belovsky, O., Shirazi, M., and Oertel, G. W.**, On the metabolism of 7α-(^{3}H)-5-etienic acid methyl ester in the human, *J. Steroid Biochem.*, 5, 491, 1974.
175. **Monder, C.**, unpublished observations.

176. **Belovsky, O., Benes, P., and Oertel, G. W.**, Inhibition of glucose-6-phosphate dehydrogenase by steroids. VIII. Effects of synthetic C_{19} and C_{20} steroids on placental glucose-6-phosphate dehydrogenase, *J. Steroid Biochem.*, 5, 697, 1974.
177. **Nayfeh, S. N. and Baggett, B.**, Metabolism of progesterone by rat testicular homogenates. III. Inhibitory effects of intermediates and other steroids, *Steroids*, 14, 269, 1969.
178. **Voigt, W. and Hsia, S. L.**, Further studies on testosterone 5α-reductase of human skin. Structural features of steroid inhibitors, *J. Biol. Chem.*, 248, 4280, 1973.
179. **Voigt, W. and Hsia, S. L.**, The antiandrogenic action of 4-androsten-3-one-17-β-carboxlic acid and its methyl ester on hamster flank organ, *Endocrinology*, 92, 1216, 1973.
180. **Hsia, S. L. and Voigt, W.**, Inhibition of dihydrotestosterone formation: an effective means of blocking androgen action in hamster sebaceous gland, *J. Invest. Dermatol.*, 62, 224, 1974.
181. **Nozu, K. and Tamaoke, B.**, Characteristics of the nuclear and microsomal steroid Δ^4-5α-hydrogenase of the rat prostate, *Acta Endocrinol. (Copenhagen)*, 16, 608, 1974.
182. **Singer, F. M., Januszka, J. P., and Borman, A.**, New inhibitors of in vitro conversion of acetate and mevalonate to cholesterol, *Proc. Soc. Exp. Biol. Med.*, 102, 370, 1959.
183. **Martyr, R. J. and Benisek, W. F.**, Chemical modification of amino acid residues associated with the Δ^4-3-ketosteroid-dependent photoinactivation of Δ^5-3-ketosteroid isomerase, *J. Biol. Chem.*, 250, 1218, 1975.
184. **Atkinson, R. M., Davis, B., Pratt, M. A., Sharpe, H. M., and Tomich, E. G.**, Action of some steroids on the central nervous system of the mouse. II. Pharmacology, *J. Med. Chem.*, 8, 426, 1965.
185. **Langecker, H.**, Der Plasmaspiegel des Pregnan-21-ol-3,20-dione Hemisuccinate Na (Hydroxydione) und Seiner Metabolite bei der anwendung als Narkotikem am Menschen, *Acta Endocrionol. (Copenhagen)*, 30, 369, 1959.
186. **Marusic, E. T., White, A., and Aedo, A. R.**, Oxidative reactions in the formation of an aldehyde group in the biosynthesis of aldosterone, *Arch. Biochem. Biophys.*, 157, 320, 1973.
187. **Neher, R. and Wettstein, A.**, Isolierung und Konstitutionsermittlung weiterer Pregnanverbindungen aus Nebennieren, *Helv. Chim. Acta*, 39, 2062, 1956.
188. **Ulick, S.**, Normal and alternate pathways in aldosterone biosynthesis, in *Endocrinology, Proc.* 4th Int. Congr., Endocrinology, Scow, R. O., Ed., American Elsevier, New York, 1973, 761.
189. **Kahnt, F. W. and Neher, R.**, Über die adrenale Steroidbiosynthese in vitro. I. Umwandlung endogener und exogener Vorstufen im Nebennierenhomogenat des Rindes, *Helv. Chim. Acta*, 48, 1457, 1965.
190. **Kohler, H., Hesse, R. H., and Pechet, M. M.**, The metabolism of aldosterone. Metabolic pathway isolation, characterization and synthesis of metabolites, *J. Biol. Chem.*, 239, 4117, 1964.
191. **Monder, C.**, Studies on the reaction of cytochrome C with corticosteroids, *Biochim. Biophys. Acta*, 164, 369, 1968.
192. **Teschke, R., Hasumura, Y., and Lieber, C. S.**, Hepatic ethanol metabolism; respective roles of alcohol dehydrogenase, the microsomal ethanol-oxidizing system, and catalase, *Arch. Biochim. Biophys.* 175, 635, 1976.
193. **Monder, C. and White, A.**, Conversion of cortisol 21-aldehyde to cortisol by rat tissues, *Biochim. Biophys. Acta*, 46, 410, 1961.
194. **Schneider, J. J.**, Enzymatic reduction of compound E-21-aldehyde to compound E, *J. Am. Chem. Soc.*, 75, 2024, 1953.
195. **Monder, C. and Martinson, F. R. E.**, Partial purification of a NADPH dependent 21-hydroxysteroid dehydrogenase from human placenta, *Biochim. Biophys. Acta*, 171, 217, 1964.
196. **Monder, C. and Furfine, C. S.**, 21-Hydroxysteroid dehydrogenases of liver and adrenal, in *Methods of Enzymology* Vol. 15, Clayton, R. B. and Lowenstein, J., Eds., Academic Press, New York, 1969, 667.
197. **Monder, C. and White, A.**, Purification and properties of a sheep liver 21-hydroxysteroid nicotinamide adenine dinucleotide oxidoreductase, *J. Biol. Chim.*, 238, 767, 1963.
198. **Monder, C. and White, H.**, The 21-hydroxysteroid dehydrogenases of liver, *J. Biol. Chem.*, 240, 71, 1965.
199. **Monder, C.**, unpublished data.
200. **Gerhards, E., Nieuweboer, B., Schulz, G., and Gibian, H.**, Stoffinechsel von 6α-fluor-16α-methyl-pregna-1,4-diene-11β,21-diol-3,20-dion (Fluocortolon) Beim menschen, *Acta Endocrinol. (Copenhagen)*, 68, 98, 1971.
201. **Cooke, A. M., Rogers, A. W., and Thomas, G. H.**, The urinary metabolites of progesterone labelled with tritium and ^{14}C in the rabbit, *J. Endocrinol.*, 27, 299, 1963.
202. **Thomas, G. H.**, Ketonic metabolites of progesterone and 19-norprogesterone in rabbit urine, *Biochem. J.*, 83, 450, 1962.
203. **Senciall, I. R. and Thomas, G. H.**, Urinary and biliary excretion of (14-^{14}C) 20α and (4-^{14}C) 20β-hydroxypregn-4-en-3-one metabolites in the rabbit, *J. Endocrinol.*, 48, 61, 1970.
204. **Taylor, W. and Scratcherd, T.**, Steroid metabolism in the rabbit, biliary and urinary excretion of metabolites of (4-^{14}C) progesterone, *Biochem. J.*, 97, 89, 1965.

205. **Allen, J. G., Cooke, A. M., and Thomas, G. H.**, The structure of some reaction products of the acidic urinary metabolites of progesterone, *J. Endocrinol.*, 40, 153, 1968.
206. **Senciall, I. R.**, Acidic steroid metabolites. Application of alumina adsorption chromatography in the differentiation of an acidic progesterone-^{14}C metabolite fraction in rat bile and urine, *Biochem. Med.*, 8, 423, 1973.
207. **Senciall, I. R. and Dey, A. C.**, Acidic steroid metabolites: evidence for the excretion of C-21-carboxylic acid metabolites of progesterone in rabbit urine, *J. Steroid Biochem.*, 7, 125, 1976.
208. **Dey, A. C. and Senciall, I. R.**, Acidic steroid metabolites: *in vitro* metabolism of tritiated progesterone and deoxycorticosterone by rabbit liver, *J. Steroid Biochem.*, 7, 167, 1976.
209. **Dey, A. C. and Senciall, I. R.**, *In vitro* formation of acidic metabolites of progesterone and desoxycorticosterone by rabbit liver, *Can. Fed. Biol. Soc.* 18 (Abstr. 383), 96, 1975.
210. **Senciall, I. R., Harding, C. A., and Dey, A. C.**, Acidic steroid metabolites: specific differences in the urinary excretion of acidic metabolites of progesterone, *J. Endocrinol.*, 68, 161, 1976.
211. **Bhavnani, B. R., Shah, K. N., and Solomon, S.**, *In vitro* metabolism of 15α-hydroxyprogesterone by rabbit liver, *Biochemistry*, 11, 753, 1972.
212. **Dey, A. C. and Senciall, I. R.**, C-21 and 6α-hydroxylation of progesterone by rabbit liver subcellular fractions, *Can. J. Biochem.*, 55, 602, 1977.
213. **Senciall, I. R., Harding, C. A., and Dey, A. C.**, Acidic steroid metabolites; predominant excretion of 20-oxo-pregnanoic acid metabolites of progesterone in rabbit urine. *J. Steroid Biochem.*, 9, 385, 1978.
214. **Riegel, B., Hartop, W. L., Jr., and Kittinger, G. W.**, Studies in the metabolism of radio-progesterone in mice and rats, *Endocrinology*, 47, 311, 1950.
215. **Shen, C. N.-H., Elliott, W. H., Doisy, E. A., Jr., and Doisy, E. A.**, The excretion of metabolites of progesterone-21-^{14}C after intragastric administration to rats, *J. Biol. Chem.*, 208, 133, 1954.
216. **Grady, H. J., Elliott, W. H., Doisy, E. A., Jr., Bocklage, B. C., and Doisy, E. A.**, Synthesis and metabolic studies of progesterone-21-^{14}C, *J. Biol. Chem.*, 195, 755, 1952.
217. **Gallagher, T. F. and Kritchevsky, T. H.**, Perbenzoic acid oxidation of 20-keto steroids and the stereochemistry of C-17, *Recent Progr. Horm. Res.*, 6, 142, 1951.
218. **Monder, C. and Wang, P. T.**, Oxidation of 21-dehydrocorticosteroids to steroidal 20-oxo-21-oic acids, *J. Biol. Chem.*, 248, 8547, 1973.
219. **Monder, C. and Wang, P. T.**, Oxidation of 21-dehydrocorticosteroids to steroidal 20-oxo-21-oic acids by ketoaldehyde dehydrogenase of sheep liver, *J. Steroid Biochem.*, 4, 153, 1973.
220. **Monder, C.**, Alpha-keto aldehyde dehydrogenase, an enzyme that catalyzes the enzymic oxidation of methylglyoxal to pyruvate, *J. Biol. Chem.*, 242, 9603, 1967.
221. **Urata, G. and Granick, S.**, Biosynthesis of α-aminoketones and the metabolism of aminoacetone, *J. Biol. Chem.*, 238, 811, 1963.
222. **Schull, K. H. and Miller, O. N.**, Formation *in vivo* of glycogen by certain intermediates of the lactate-propanediol pathway, *J. Biol. Chem.*, 235, 551, 1960.
223. **Teng, S. M., Sellinger, O. Z., and Miller, O. M.**, The metabolism of lactaldehyde. VI. The reduction of D- and L-lactaldehyde in rat liver, *Biochim. Biophys. Acta*, 89, 217, 1964.
224. **Lee, H. J. and Monder, C.**, Oxidation of corticosteroids to steroidal carboxylic acids by an enzyme preparation from hamster liver, *Biochemistry*, 16, 3810, 1977.
225. **Martin, K. O., Oh, S. W., Lee, H-J., and Monder, C.**, Studies of 21-^{3}H labeled corticosteroids: evidence for isomerization of the ketol side chain of 11-deoxycorticosterone by a hamster liver enzyme, *Biochemistry*, 16, 3803, 1977.
226. **Lewbart, M.**, private communication.
227. **Monder, C.**, unpublished observations.
228. **Monder, C. and Oh, C. W.**, in preparation.
229. **Franzen, V.**, Disproportionation of methylglyoxal catalyzed by cyanide, *Chem. Ber.*, 89, 2154, 1956.
230. **Meyerhof, O.**, Observations concerning methylglyoxalase, *Biochem. Z.*, 159, 432, 1925.
231. **Monder, C.**, Synthesis of steroids containing the α-keto acid side chain, *Steroids*, 18, 187, 1971.
232. **Lewbart, M. L. and Mattox, V. R.**, Conversion of steroid-17-yl-glyoxals to epimeric glycolic esters, *J. Org. Chem.*, 28, 1779, 1963.
233. **Laurent, H., Gerhards, E., and Wiechert, R.**, New biologically active pregnan-21-oic acid esters, *Angw. Chem. Int. Ed. Eng.*, 14, 65, 1975.
234. **Corey, E. J., Gilman, N. W., and Ganen, B. E.**, New methods for the oxidation of aldehydes to carboxylic acids and esters, *J. Chem. Soc.*, 90, 5616, 1968.
235. **Silber, R. H. and Porter, C. C.**, The determination of 17,21-dihydroxy-20-ketosteroids in urine and plasma, *J. Biol. Chem.*, 210, 923, 1954.
236. **Lewbart, M.**, Formation and Reactions of Steroidal Glyoxals. I. Reactions of α-Ketolic Steroids with Methanolic Cupric Acetate. II. The Mechanism of the Porter-Silber Reaction, Ph.D. thesis., University of Minnesota, Minneapolis, 1961.

237. **Nef, J. U.**, Uber das Verhalten des Acetols bzw. Benzoylcarbinols gegen fehlingsche und andere Oxydationsmittel, *Annalen der Chemie,* 335, 269, 1904.
238. **Evans, W. L.**, On the behavior of benzoyl carbinol towards alkalies and oxidizing agents, *Am. Chem. J.*, 35, 115, 1906.
239. **Lewbart, M. L.**, Preparation and properties of β-lactones from steroidal 17,20-dihydroxy-21-oic acids, *J. Org. Chem.*, 37, 1224, 1972.
240. **Etienne, Y. and Fischer, N.**, β-Lactones, *Chem. Heterocycl. Compd.*, 19(2), 796, 1964.
241. **Lewbart, M. L. and Schneider, J. J.**, Preparation and properties of steroidal 17,20- and 20,21-acetonides epimeric at C-20. II. Derivatives of cortisol and cortisone, *J. Org. Chem.*, 34, 3513, 1969.
242. **Lewbart, M. and Schneider, J. J.**, Preparation and properties of steroidal 17,20 and 20,21 acetonides epimeric at C-20. I. Derivatives of 5β-pregnan-3α-ol, *J. Org. Chem.*, 34, 3505, 1969.
243. **Lewbart, M.**, Preparation and properties of steroidal 17,20- and 20,21-acetonides epimeric at C-20. III. Dioxolone derivatives of α-hydroxy acids, *J. Org. Chem.*, 36, 586, 1971.
244. **Benn, W. R.**, A reinvestigation of the synthesis of 5,16 pregnadiene-3β,20α-diol. NMR spectra of some epimeric 20 hydroxypregnane derivatives, *J. Org. Chem.*, 28, 3557, 1963.
245. **Robinson, C. H. and Hofer, P.**, NMR Spectra of C-20-substituted pregnanes, *Chem. Ind. (N.Y.),* 377, 1966.
246. **Fieser, L. F. and Fieser, M.**, *Steroids,* Reinhold, New York, 1959.
247. **Lewbart, M. L. and Mattox, V.**, Conversion of steroid-17-yl glyoxals to epimeric glycolic esters, *J. Org. Chem.*, 28, 1779, 1963.
248. **Lewbart, M. L. and Schneider, J. J.**, Preparation of six 20-deoxy steroids in the 3α-hydroxy-5β-pregnane series and their use in optical rotation studies, *J. Org. Chem.*, 33, 1707, 1968.
249. **Lewbart, M. L. and Schneider, J. J.**, Glycolic acid metabolites of desoxycorticosterone assignment of configuration at C-20, *J. Org. Chem.*, 29, 2559, 1964.
250. **Bradlow, H. L., Zumoff, B., Monder, C., Lee, H.-J., and Hellman, L.**, Isolation and identification of four new carboxylic acid metabolites of cortisol in man, *J. Clin. Endocrinol. Metab.*, 37, 811, 1973.
251. **Willingham, A. K. and Monder, C.**, Evidence for the metabolic lability of the C-21 hydrogens of corticosteroids, *Endocr. Res. Commun.*, 1, 145, 1974.
252. **Lee, H. J.**, Acidic metabolite of prednisolone, *Experientia,* 33, 253, 1977.
253. **Martin, K. and Monder, C.**, Oxidation of corticosteroids to steroidal-21-oic acids by human liver enzyme, *Biochemistry,* 15, 576, 1976.
254. **Bradlow, H. L., Zumoff, B., Monder, C., and Hellman, L.**, C-21 oxidation of cortisol in man, *J. Clin. Endocrinol. Metab.*, 37, 805, 1973.
255. **Kendall, E. C., Mason, H. L., McKenzie, B. F., Myers, C. S., Koelsche, G. A.**, Isolation in crystalline form of the hormone essential to life from the suprarenal cortex, *Proc. Staff Meet. Mayo Clin.*, 9, 245, 1934.
256. **Oh, S. and Monder, C.**, Synthesis of corticosteriod derivatives containing the 20β-ol-21-al side chain, *J. Org. Chem.*, 41, 2477, 1976.
257. **Lippman, V. and Monder, C.**, Enzyme mediated reduction of 21-dehydrocorticosteroids at C-20, *J. Steroid Biochem.*, 7, 719, 1976.
258. **Monder, C. and Bradlow, H. L.**, General review of carboxylic acid metabolites of steroids, *J. Steroid Biochem.*, 8, 897, 1977.
259. **Fukushima, D. K., Bradlow, H. L., Hellman, L., Zumoff, B., and Gallagher, T. F.**, Metabolic transformation of hydrocortisone-4-^{14}C, *J. Biol. Chem.*, 235, 2246, 1960.
260. **Martin, K. O. and Monder, C.**, manuscript in preparation.
261. **Greenfield, N. J. and Pietruazka, R.**, Two aldehyde dehydrogenases from human liver. Isolation via affinity chromatography and characterization of the isozymes, *Biochim. Biophys. Acta,* 483, 35, 1977.
262. **Eckfeldt, J., Mope, L., Takio, K., and Yonetani, T.**, Horse liver aldehyde dehydrogenase. Purification and characterization of two iso-enzymes, *J. Biol. Chem.*, 251, 236, 1976.
263. **Lippman, V. and Monder, C.**, Purification and properties of an NADPH dependent 17 aldol reductase from sheep liver, *J. Biol. Chem.*, 253, 2126, 1978.
264. **Rose, I. A.**, Mechanism of the aldose-ketose isomerase reactions, *Adv. in Enzymol. Rela. Subj. Biochem.*, 43, 491, 1975.
265. **Cahn, R. S., Ingold, C. K., and Prelog, V.**, The specification of asymmetric configuration in organic chemistry, *Experientia,* 12, 81, 1956.
266. **Orr, J. C. and Monder, C.**, The stereospecificity of the enzymic reduction of 21-dehydrocortisol, *J. Biol. Chem.*, 250, 7547, 1975.
267. **Willingham, A. K. and Monder, C.**, Synthesis of 21-^{3}H corticosteroids, *Steroids,* 22, 539, 1973.
268. **Bradlow, H. L., Zumoff, B., Monder, C., and Hellman, L.**, Cortisol → cortoic acids metabolic pathways in man, *J. Steroid Biochem.*, 5, 323, 1974.

269. **Zumoff, B., Monder, C. and Bradlow, L.,** Studies in the biotransformation of cortisol to the cortoic acids in man. IV. The central role of tetrahydrocortisol and tetrahydrocortisone as intermediates, *J. Clin. Endocrinol. Metab.,* 44, 647, 1977.
270. **Weiss, G., Monder, C., and Bradlow, L.,** 17-Deoxygenation, a new pathway of cortisol metabolism isolation of 17-deoxy cortolonic acids, *J. Clin. Endocrinol. Metab.,* 43, 696, 1976.
271. **Mattox, V. R.,** The formation of a glyoxal side chain at C-17 from steroids with dihydroxyacetone and Δ^{16}-ketol side chains, *J. Am. Chem. Soc.,* 74, 4340, 1952.
272. **Durst, H. D., Milano, M., Kikta, E. J., Jr., Connelly, S. A., and Grushka, E.,** Phenacyl esters of fatty acids via crown ether catalysts for enchanced ultraviolet detection in liquid chromatography, *Anal. Chem.,* 47, 1797, 1975.
273. **Farhi, R. L. and Monder, C.,** Analysis of steroidal carboxylic acids by high pressure chromatography, *Anal. Biochem.,* 90, 58, 1978.
274. **Senciall, I. R., Riley, C., Rahal, S., Dey, A. C., Harding, C. A., and Acharya, M.,** Effect of structure on the excretion of steroidal acids in human urine and their potential colorimetric assay as 17-oxogenic steroids, *Clin. Biochem.,* 10, 32, 1977.
275. **Bradlow, H., Monder, C., and Zumoff, B.,** Studies in the biotransformation of cortisol to cortoic acids in man. III. 21-Oxidation of 4-^{14}C 21-^{3}H-desoxycorticosterone, *J. Clin. Endocrinol. Metab.,* 45, 960, 1977.

Chapter 3

THE BIOCHEMISTRY OF THE 17-HYDROXYSTEROID DEHYDROGENASES

D. G. Williamson

TABLE OF CONTENTS

I. INTRODUCTION

The 17-hydroxysteroid dehydrogenases (17-oxidoreductases) are pyridine nucleotide-dependent enzymes that catalyze the stereospecific interconversion of the hydroxyl and carbonyl groups at C-17 of the C_{18} and C_{19} steroids. Since chemical modification at C-17 of estrogens and androgens can affect their biological activity, the 17-hydroxysteroid dehydrogenases have a potentially important role in regulating the physiological action of estrogenic and androgenic steroids. Consequently, considerable attention has been directed to these enzymes with regard to their chemical and physical properties and function in mammalian tissues. The majority of enzymes involved in the biosynthesis of steriod hormones are associated with the membrane structures of cells[1] which precludes enzyme purification and characterization. However, a number of 17-hydroxysteroid dehydrogenases are located in the cytosol of mammalian cells, and some have been isolated in a homogeneous state. The soluble 17β-hydroxysteroid dehydrogenase of human placenta, first characterized by Langer and Engel[2] in 1958, is the most extensively studied mammalian 17-hydroxysteroid dehydrogenase. It has been both purified by several groups employing a variety of chromatographic techniques[3-7] and crystallized.[8] The chemical and physical properties of this placental enzyme have been recently summarized.[9]

This chapter will describe a number of the 17-hydroxysteroid dehydrogenases in mammalian tissues and discuss the current knowledge with regard to the properties, regulation, and biological function of these enzymes. Most of the data presented in this chapter are the results of studies carried out in the author's laboratory on the 17-hydroxysteroid dehydrogenases of the rabbit, the properties of which are common to those of many other mammalian species.

II. GENERAL PROPERTIES OF THE 17-HYDROXYSTEROID DEHYDROGENASES

The 17-hydroxysteroid dehydrogenases consist of the 17β and 17α enzyme groups. This division is based on the stereospecificity of the enzymatic reaction at C-17 of the steroid substrate; 17β-hydroxysteroid dehydrogenases catalyze the interconversion of the 17-keto- and 17β-hydroxysteroids while 17α-hydroxysteroid dehydrogenases catalyze the interconversion of 17-keto- and 17α-hydroxysteroids. 17β-Hydroxysteroid dehydrogenase activity is present in all mammalian species, however, 17α-hydroxysteroid dehydrogenase activity is found primarily in species which excrete the 17α-epimers of estrogens and androgens. For example, in ruminants and the rabbit,[10] 17α-estradiol is the major urinary metabolite of 17β-estradiol. The transformation of 17β-estradiol to 17α-estradiol has been demonstrated in the rabbit both in vivo[11] and in vitro.[12] Using 17β-estradiol labeled with tritium at the 17α-position, Williams et al.[13] showed that the in vivo conversion of this compound to 17α-estradiol proceeded via estrone and not by epimerization. Thus, in this species, the metabolism of 17β-estradiol involves both the 17β- and α-hydroxysteroid dehydrogenases. In general, the conversion of the 17β-epimers of the C_{18} and C_{19} steroids to the 17α-epimers follows the reaction sequence demonstrated with the rabbit, although Szamatowicz et al.[14] has suggested (from in vitro studies with guinea pig ovaries and testes) that direct conversion of testosterone to epitestosterone by epimerization could occur.

Purification of the 17-hydroxysteroid dehydrogenases from different species indicates that the 17β- and α-hydroxysteroid dehydrogenase activities are dependent upon distinct enzymes. Ball and Breuer[15] reported that a single enzyme was responsible for both activities in rabbit liver cytosol based on kinetic data and a number of purification procedures. However, Hasnain and Williamson[16,17] were able to separate the 17β- and α-hydroxysteroid dehydrogenase activities of rabbit liver cytosol by DEAE- cellulose chromatography. Renwick and Engel[18] have achieved a partial separation of the soluble 17β- and α-hydroxysteroid dehydrogenase of chicken liver.

A. Tissue Distribution

The presence of 17-hydroxysteroid dehydrogenase activity in a variety of mammalian tissues has been well documented. 17β-Hydroxysteroid dehydrogenase activity is not restricted to the ovary,[18] testes,[20] and placenta[2] or to hormone-responsive tissues such as the uterus[21] or ventral prostate.[22] Enzyme activity has been demonstrated in such tissues as liver, kidney, skin, and blood of several species;[23-25] additionally, it has been shown to occur in certain bacteria.[26,27] 17α-Hydroxysteroid dehydrogenase activity is present primarily in the liver and kidney of various species.[14,18,28-30] However, activity has been found in mare ovary[31] and placenta,[32] human blood and adrenal tissue,[33] and ox and sheep blood.[34]

The pyridine nucleotide specificity of the 17-hydroxysteroid dehydrogenases can vary among tissues in the same species and among the same tissues in different species; the 17-hydroxysteroid dehydrogenases of rabbit tissues are typical examples of this variation. In our laboratory, the tissue distribution of the NAD- and NADP-dependent 17-hydroxysteroid dehydrogenases have been examined in the rabbit (Table 1). The 17β enzyme activity was present in the homogenates of all tissues examined and highest in liver homogenates. Significant 17α activity is present only in liver and kidney homogenates. With regard to pyridine nucleotide specificity, the 17β activities of the liver and large intestine exhibit a preference for NAD, while NADP is the preferred cofactor in the kidney and small intestine. NADP is also the preferred cofactor for the 17α enzyme activities of liver and kidney.

TABLE 1

Tissue Distribution of the NAD- and NADP-Dependent 17-Hydroxysteroid Dehydrogenases in the Rabbit

Tissue	Enzyme activity (pmol product/min/mg protein)[a]			
	17β-Estradiol		17α-Estradiol	
	NAD	NADP	NAD	NADP
Liver	350	60	0.5	9.6
Kidney	24	39	8.4	17
Small intestine	4.0	8.5	0.0	0.04
Large intestine	41	16	0.0	0.07
Spleen	1.5	1.8	0.03	0.03
Uterus	0.12	0.30	0.01	0.02

Note: 17β- and α-hydroxysteroid dehydrogenase activities were assayed with the substrates 17β- and α-estradiol, respectively.

[a] A unit of enzyme activity is defined as the amount of enzyme catalyzing the oxidation of 1 μmol of substrate per minute at 37°C.

B. Subcellular Distribution

Studies on the subcellular distribution of the 17-hydroxysteroid dehydrogenases of mammalian systems have shown that these enzymes are located in the particulate fractions of some tissues and are soluble enzymes in other tissues. Some tissues have both soluble and particulate activities. Such is the case in rabbit liver, where 17β-hydroxysteroid dehydrogenase activity toward androgens and estrogens is present in both the microsomal and soluble fractions (Table 2). The microsomal enzyme activity is essentially specific for NAD; however, the soluble enzyme can utilize either NAD or NADP. It is interesting to note that soluble 17β enzyme activity toward 17β-estradiol is higher with NAD while activity toward testosterone is higher with NADP. The 17α-hydroxysteroid dehydrogenase activity of rabbit liver is found essentially only in the soluble fraction of the cell and has a marked preference for NADP.

The properties of the 17-hydroxysteroid dehydrogenases of rabbit liver with regard to subcellular distribution and nucleotide specificity (Table 2) are representative of many of the 17-hydroxysteroid dehydrogenases of other species and tissues. A summary of the properties of a number of these enzymes is presented in Table 3. This table does not list all of the enzymes that have been studied; rather, it demonstrates the variation in subcellular distribution and pyridine nucleotide specificity that can occur. Although it is difficult to generalize, NAD is the preferred cofactor in the majority of tissues for the 17β-hydroxysteroid dehydrogenases located in the microsomal fraction and NADP is the cofactor for the soluble enzyme activity. However, Aoshima and Kochakian[23] examined the 17β enzymes of the liver and kidney of the guinea pig, rabbit, mouse, rat, hamster, and dog and found that, in some of these species, the NAD-dependent enzyme was differentially distributed between the microsomal and soluble fractions. Although the 17β enzymes that utilize NADP are usually restricted to the soluble fraction of tissues, in the testes of the pig[20] and other species[44] this enzyme activity is localized in the microsomal fraction. The 17α-hydroxysteroid dehydrogenases are primarily soluble enzymes and utilize NADP as cofactor (Tables 2 and 3). Two exceptions are the soluble enzymes of mouse kidney and horse placenta, these are more effective with NAD.[30,32]

TABLE 2

Subcellular Distribution of the NAD- and NADP-Dependent 17-Hydroxysteroid Dehydrogenases of Rabbit Liver

Substrate	Enzyme activity (pmol product/min/mg protein)[a]			
	NAD		NADP	
	Cytosol	Microsomes	Cytosol	Microsomes
Testosterone	895	626	6550	7.7
17β-Estradiol-	336	365	126	15
Epitestosterone	242	4.6	2192	31
17α-Estradiol	18	0.08	114	1.5

[a] A unit of enzyme activity is defined as the amount of enzyme catalyzing the oxidation of 1 μmol of substrate per minute at 37°C.

TABLE 3

Subcellular Distribution and Pyridine Nucleotide Specificity of the 17-Hydroxysteroid Dehydrogenases

Enzyme source	Subcellular location	Pyridine nucleotide specificity	Ref.
	17β-Hydroxysteroid Dehydrogenases		
Pig testis	Microsomes	NADP	20
Sheep ovary	Cytosol	NAD	19, 35
Dog prostate	Microsomes	NAD	22, 36
Human endometrium	Mitochondria	NAD	37, 38
	Microsomes	NAD	37, 39
	Cytosol	NAD	37, 40
Human placenta	Cytosol	NAD ⩾ NADP	4
Human erythrocytes	Hemolysate	NADP	25
Rat skin	Microsomes	NAD	24
Guinea pig liver	Microsomes	NAD	23, 41
	Cytosol	NADP	23, 41
Guinea pig kidney	Cytosol	NADP	23, 42
Mouse kidney	Cytosol	NADP	30
	Microsomes	NAD	30
Human liver	Microsomes	NAD	43
	Cytosol	NADP	43
	17α-Hydroxysteroid Dehydrogenases		
Mouse kidney	Cytosol	NAD	30
Chicken liver	Cytosol	NADP	18
Sheep liver	Cytosol	NADP	18

III. 17β-HYDROXYSTEROID DEHYDROGENASES

A. Human Placenta

The presence of a 17β-hydroxysteroid dehydrogenase in the soluble fraction of human placenta was first reported by Langer and Engel.[2] This enzyme has been purified by a variety of techniques[3-7] in quantities sufficient to allow detailed studies of its properties. Either NAD or NADP can function as cofactor. Karavolas et al.[4] determined the kinetic constants for these pyridine nucleotides, and the values for maximal velocity obtained with NADP were 70 to 80% of those obtained with NAD. A variety of steroids have been examined as substrates for the purified 17β-hydroxysteroid dehydrogenase.[9] The enzyme has an absolute stereochemical specificity for the 17β-hydroxyl group. The most reactive substrates are those which have an aromatic A or B ring, and the relative rate of testosterone oxidation is less than 1% that of 17β-estradiol. Indeed, Karavolas and Engel[45] reported that there was no reduction of NAD when either testosterone or 19-nortestosterone was incubated with the purified 17β enzyme. Engel and Groman[9] found that, although estriol was not a substrate, 16α-hydroxyestrone could be readily reduced to estriol, suggesting that one of the physiological roles of the placental 17β enzyme occurs in the final step of estriol synthesis. In similar studies, Findlay and Breuer[46] found that the partially purified 17β-enzyme could not catalyze the oxidation of 18-hydroxyestradiol-17β although it could catalyze the reduction of 18-hydroxyestrone.

Lehmann and Breuer[47] first observed a membrane-bound 17β-hydroxysteroid dehydrogenase in human placenta. Studies by Pollow et al.[48] indicated that the enzyme activity was present in all subcellular fractions except nuclei and that the mitochondrial activity was associated with the outer membrane of this organelle. However, Thomas and Veerkamp[49] recently showed that the buffer medium composition had a direct influence on the enzyme distribution pattern during subcellular fractionation. They concluded that the particulate 17β-hydroxysteroid dehydrogenase was located only in the microsomal fraction. There is some controversy with regard to the steroid specificity of the particulate enzyme. Isurugi et al.[50] observed an absolute specificity for testosterone with their particulate enzyme; however, Thomas and Veerkamp[49] found the highest specific activity with 17β-estradiol. NAD is the preferred cofactor of the microsomal 17β enzyme.[47,49]

B. Uterus and Ovary

The 17β-hydroxysteroid dehydrogenases of uterine tissues have been studied primarily to establish the role of these enzymes in the biological action of estrogens on the uterus. The oxidation of 17β-estradiol to estrone by human endometrium was first demonstrated by Ryan and Engel.[21] The 17β enzyme activity in this tissue has been shown to be 40-fold higher than the enzyme activity in myometrium.[51] Moreover, in human endometrial tissue, 17β activity varies during the menstrual cycle. Tseng and Gurpide[52] demonstrated that the enzyme activity in crude homogenates of endometrium increased during the secretory phase. These observations have been confirmed by Pollow et al.[37] Using histochemical techniques, Scublinsky et al.[53] found that the 17β enzyme of human endometrium was localized in the glandular epithelium of secretory endometrium; no activity was detected in proliferative tissue. A similar dependence of the uterine 17β-hydroxysteroid dehydrogenase on the estrous cycle in adult female rats has been reported by Wenzel et al.[54] Maximum oxidation of 17β-estradiol to estrone was found at the first day of estrus.

The subcellular distribution of the 17β-hydroxysteroid dehydrogenase of human endometrium has been examined by Pollow et al.[37] Activity was located mainly in the mitochondrial and microsomal fractions, and the specific activity in these fractions was 20 to 50 times greater than that in nuclei or cytosol. In further studies, Pollow et al.[38-40] characterized the 17β-hydroxysteroid dehydrogenase of the mitochondrial, microsomal, and cytosol fractions of human endometrium. They found that the microsomal enzyme was bound tightly to the membrane while that in the mitochondra associated mainly with the outer membrane fraction. The kinetic properties of the enzyme activities in the three subcellular fractions were very similar. NAD was the preferred cofactor with all three enzymes, and the interconversion of 17β-estradiol and estrone proceeded at a more rapid rate than did that of testostrone and androstenedione. One exception, however, was found in incubations carried out with the mitochondrial fraction obtained from endometrial carcinoma tissue; in this case, androstenedione was reduced more rapidly than estrone.[38]

The most detailed studies on the properties of ovarian 17β-hydroxysteroid dehydrogenases have been carried out by Kautsky and Hagerman[19] and Michel et al.[35] on partially purified enzyme from sheep ovaries. The enzyme was found only in the soluble fraction of ovary homogenates and was relatively specific for estrogens, the rate of testosterone oxidation being only 1% that observed with 17β-estradiol. Michel et al.[35] examined a number of substrate analogues and found that substituents at C_3, C_{11}, or C_{13} did not greatly affect the reaction rates with these steroids. The ovarian 17β-hydroxysteroid dehydrogenase exhibited a preference for the cofactor NAD, thus, it differs from the ovarian enzymes of the rabbit[55] and immature mouse,[56] which show a preference for NADP.

C. Prostate and Testis

The 17β-hydroxysteroid dehydrogenase of dog prostate has been partially purified by Hussein and Kochakian.[22] The activity localized in the microsomal fraction was solubilized by treatment with sodium deoxycholate. Although both NAD and NADP were cofactors for the enzyme, NAD was more effective. Moreover, no separation of the NAD- and NADP-dependent enzyme activities was observed during purification. Maximal activity was observed with testosterone; androsterone and dehydroepiandrosterone were not substrates for the enzyme.

The subcellular distribution of testicular 17β-hydroxysteroid dehydrogenase activity has been examined in a number of species, and the findings have been summarized by Tamaoki and Inano.[44] In all species studied, the majority of enzyme activity was located in the microsomal fraction. The association of the testicular 17β enzymes with biomembranes has made purification of these enzymes difficult. However, Inano and Tamaoki[20] successfully solubilized the enzyme from the microsomal fraction of pig testis by sonication and subsequently purified the enzyme to homogeneity. They also studied the steroid and pyridine nucleotide specificity of the purified enzyme.[20,57] NADP(H) was the preferred coenzyme and, in the presence of the reduced coenzyme, the reduction of C-17 of androstenedione, dehydroepiandrostenedione, and estrone proceeded at similar rates. Similarly, both testosterone and 17β-estradiol were oxidized by the enzyme in the presence of NADP. The 17β-hydroxysteroid dehydrogenase also exhibited 20α-hydroxysteroid dehydrogenase activity to the extent of one seventh of the 17β enzyme activity.

D. Liver

A detailed examination of the subcellular distribution and nucleotide specificity of the 17β-hydroxysteroid dehydrogenases of the liver in a number of mammalian species has been carried out by Aoshima and Kochakian.[23] They observed a considerable variation in the distribution of the NAD-dependent enzyme activity between the microsomal and cytosol fractions among the species. NADP-dependent activity, however, was found only in the soluble fractions prepared from liver homogenates. Several of the 17β-hydroxysteroid dehydrogenases of liver have been isolated and that in the cytosol of guinea pig liver has been partially purified by Villee and Spencer[58], Endahl and Kochakian,[59] and Joshi et al.[60] This last group of researchers also carried out extensive studies on the substrate specificity of the enzyme. Both C_{18} and C_{19} steroids were substrates. With the latter, oxidation at C-17 was not greatly affected by alterations in ring A of the steroid or by the nature of the ring A/B fusion. The partially purified 17β-enzyme also exhibited 3-hydroxysteroid dehydrogenase activity, provided the hydroxyl group at C-3 was β-oriented and the ring A/B fusion was *trans*. NADP was the cofactor for both the 17β- and 3β-hydroxysteroid dehydrogenase activities. More recently, Kageura and Toki[61] separated two 17β enzyme activities from the soluble fraction of guinea pig liver. Both could utilize either NADP or NAD, although the activity observed with NADP was fivefold higher than with NAD. Although each catalyzed the oxidation of testosterone, only one of the enzymes catalyzed the oxidation of 3-hydroxyhexobarbital. It was conceivable that a single protein was responsible for the oxidation of both the steroid and drug.

Thaler-Dao et al.[62] resolved three 17β-hydroxysteroid dehydrogenase activities from rabbit liver cytosol by DEAE-cellulose chromatography and partially purified two of these enzymes. One of the enzyme activities utilized either NAD or NADP as cofactor and catalyzed the oxidation of both testosterone and 17β-estradiol. Another required NADP and had a much higher activity toward testosterone than toward 17β-estradiol. This enzyme fraction also possessed 3α-hydroxysteroid dehydrogenase activity, and to a lesser extent, 3β activity with substrates of both the 5α- and β-androstane series. It was suggested that a single protein was responsible for both the 17β- and 3-hydroxysteroid dehydrogenase activities.

E. Kidney

The majority of the 17β-hydroxysteroid dehydrogenases in kidney are located in the cytosol of this tissue.[23] A microsomal NAD-dependent enzyme activity has been observed in mouse kidney[30] in addition to a soluble NADP-dependent 17β-hydroxysteroid dehydrogenase. The soluble 17β enzyme of guinea pig kidney is the best-characterized enzyme. Purification of the enzyme activity by Liu and Kochakian[42,63] indicated the presence of several soluble 17β-hydroxysteroid dehydrogenases in this tissue; more recently, Stevenson and Kochakian[64] completed the purification of these enzyme activities. All of the enzymes exhibited dual pyridine nucleotide specificity although higher activity was observed with NADP than with NAD. With steroid substrates, highest activity was obtained with testosterone, but the 5α- and β-reduced C_{19} steroids were also substrates. Only a trace of enzyme activity was observed with 17β-estradiol.

F. Other Tissues

17β-hydroxysteroid dehydrogenase activity toward both estrogens and androgens has been demonstrated in the blood of several mammalian species. Van der Molen and Groen[65] observed enzyme activity catalyzing the interconversion of androstenedione and testosterone in the peripheral venous blood of the human, sheep, dog, and rabbit. The activities were located in the RBC of heparinized blood. A 17β enzyme catalyzing the interconversion of 17β-estradiol and estrone has been detected in the peripheral blood of pregnant women early in gestation.[66] The serum levels of this enzyme increase severalfold during pregnancy. A similar enzyme activity has been observed in the serum of human umbilical cord blood and is elaborated by both the placenta and fetus.[67] A 17β-hydroxysteroid dehydrogenase of human erythrocytes has been partially purified and characterized by Jacobsohn and Hochberg[25] and by Mulder et al.[68] It was reported that the enzyme could be isolated from the membrane-free hemolysate of erythrocytes and had an almost absolute requirement for NADP. Thus, the human enzyme differs from the enzyme of sheep erythrocytes, which utilizes NAD,[69] and the enzyme of rat erythrocytes, which uses either coenzyme.[70] Studies on the human erythrocyte 17β enzyme have shown that NADP, in addition to its role as coenzyme, prevents enzyme inactivation.[71] Indeed, this pyridine nucleotide activates the enzyme upon binding. The 17β-hydroxysteroid dehydrogenase of human erythrocytes has a broad substrate specificity and catalyzes the conversion of both androgens and estrogens with a 17β-hydroxyl group or 17-oxo group. In addition, the sulfate derivatives of estrone, 17β-estradiol, and dehydroepiandrosterone are also substrates.[25,68]

Specific areas of the brain such as the hypothalamus and pituitary can concentrate 17β-estradiol and testosterone.[72,73] Moreover, these steroid hormones can influence the secretion of gonadotropins from the pituitary.[74] Consequently, the involvement of 17β-hydroxysteroid dehydrogenases in the metabolism and function of these steroids in central nervous tissue has been widely examined. In vivo reduction of estrone to 17β-estradiol and in vitro oxidation of testosterone to androstenedione by rat central nervous tissues has been reported.[75-77] 17β-Hydroxysteroid dehydrogenase activity has also been observed in the central nervous tissues of human fetus, monkeys, and dogs.[78,79] Reddy et al.[80] found that in the rabbit CNS the adenohypohysis was the major site of 17β-hydroxysteroid dehydrogenase activity when compared to the hypothalamus, limbic system, and cortex. In contrast, no significant differences in the specific activity of this enzyme in the brain tissues of the male rat were noted by Rommerts and Van der Molen.[81] Using subcellular fractionation studies, they also showed that the 17β-hydroxysteroid dehydrogenase of rat brain tissues was a soluble enzyme.

The influence of estrogens on biological activities in the skin has prompted investigation of steroid metabolism in this tissue.[24,82] Davis et al.[24] have examined the properties of the 17β-hydroxysteroid dehydrogenase in the skin of newborn rats. The enzyme was located in the microsomal fraction of this tissue and showed a preference for NAD as cofactor. The enzyme preparation catalyzed the oxidation of both 17β-estradiol and testosterone. The inhibitory effects of a number of C_{18} and C_{19} steroids indicated that a common structural feature of the effective inhibitors was a 17β-hydroxyl group; the structure in ring A or the presence of the methyl group at C-10 was not essential.

IV. 17α-HYDROXYSTEROID DEHYDROGENASES

In general, steroid metabolic pathways involving the 17α-hydroxysteroid dehydrogenases are those which result in inactivation and excretion of the C_{18} and C_{19} steroid hormones. In ruminants, the rabbit, and the dog, 17β-estradiol is converted to the 17α-epimer prior to excretion in the urine.[10] Epitestosterone has been identified in human urine by several researchers.[83-86]

Although 17α-hydroxysteroid dehydrogenase activity has been demonstrated in a number of different tissues,[30-34] liver and kidney are the best sources of this enzyme activity. Therefore, most studies on the properties of the 17α enzymes have been carried out with these tissues. Renwick and Engel[18] achieved a partial purification of the 17α-hydroxysteroid dehydrogenase of chicken liver and a partial separation of this enzyme activity from the 17β-hydroxysteroid dehydrogenase present in the same subcellular fraction. The 17α enzyme was located in the cytosol, required NADP as a cofactor, and catalyzed the oxidation of both 17α-estradiol and epitestosterone. A similar enzyme activity was observed in the liver homogenates of turkey and sheep. In our laboratory, we have been examining the properties of the 17α-hydroxysteroid dehydrogenases of rabbit liver and kidney. A comparison of the soluble NADP-dependent 17α- and 17β-hydroxysteroid dehydrogenase activities of these tissues is illustrated in Table 4. In liver cytosol, the oxidation of testosterone and 17β-estradiol proceeds at a more rapid rate than does the oxidation of epitestosterone or 17α-estradiol. This is particularly evident with the androgen substrates, e.g., testosterone is oxidized at four times the rate of epitestosterone. The reverse is observed in the kidney cytosol, where enzyme

TABLE 4

Substrate Specificities of the Soluble NADP-Dependent 17-Hydroxysteroid Dehydrogenases of Rabbit Liver and Kidney

Substrate	Enzyme activity (pmol product/ min/mg protein)[a]	
	Liver	Kidney
Testosterone	2710	29
Epitestosterone	788	825
17β-Estradiol	57	20
17α-Estradiol	39	41
17β-Estradiol 3-glucuronide	24	14
17α-Estradiol 3-glucuronide	132	489

[a] A unit of enzyme activity is defined as the amount of enzyme catalyzing the oxidation of 1 μmol of substrate per minute at 37°C.

activity toward epitestosterone and 17α-estradiol is higher than the activity toward the corresponding 17β-epimers. With the estrogen substrates, conjugation with glucuronic acid at C-3 of 17α-estradiol produces an increase in the rate of steroid oxidation. This is particularly evident in the kidney, where activity toward 17α-estradiol 3-glucuronide is ten times higher than that toward 17α-estradiol. This increased activity is not observed with the 3-glucuronide derivative of 17β-estradiol; indeed, in the liver cytosol, there is a significant decrease in the rate of oxidation of this substrate compared to 17β-estradiol. Purification of the 17α-hydroxysteroid dehydrogenase from the cytosol of rabbit liver has been achieved in our laboratory by a combination of DEAE-cellulose chromatography and isoelectric focusing.[16,17] Multiple forms of this enzyme activity were observed, and five of these forms were obtained in a homogeneous state. Substrate-specificity studies revealed differences in enzyme activity toward androgens and estrogens among the different enzyme forms.[87] Activity toward the androgen, epitestosterone, was higher with five of the six enzymes examined. One enzyme was 30 times more active toward epitestosterone than toward the estrogens. Of special interest was the enzyme which exhibited a greater activity toward the estrogen 17α-estradiol 3-glucuronide than toward epitestosterone. This high activity was lost if the glucuronic acid moiety was removed or replaced by glucose or galacturonic acid. This 17α-hydroxysteroid dehydrogenase, specific for the glucuronide derivative of 17α-estradiol, may play a special role in the metabolism of estrogens in the rabbit. The possible significance of this enzyme will be discussed in Section VII. B.

V. PHYSICAL AND CHEMICAL PROPERTIES OF THE 17-HYDROXYSTEROID DEHYDROGENASES

A. Molecular Weight and Subunit Structure

Of the mammalian 17-hydroxysteroid dehydrogenases, the soluble 17β-hydroxysteroid dehydrogenase of human placenta has been studied most extensively in terms of its physical properties. A variety of values for the molecular weight of this enzyme have been obtained by different methods.[9] For example, molecular weights of 62,000, 62 to 65,000, and 67.7 to 69,000 have been obtained for the native enzyme by density-gradient centrifugation,[80] gel filtration;[89] and ultrancentrifugation,[5,90] respectively. In the presence of sodium dodecyl sulfate, polyacrylamide disc gel electrophoresis has yielded values of 33,500[90] and 37,000.[91] Jarabak and Street[91] treated the native enzyme with the cross-linking agent diethyl suberimidate, subjected it to polyacrylamide disc gel electrophoresis in the presence of sodium dodecyl sulfate, and obtained a 73,000 mol wt. Thus, the placental 17β-hydroxysteroid dehydrogenase having a molecular weight of approximately 68,000 is a dimer consisting of subunits with molecular weights of 34,000. Further studies on the subunit structure of the enzyme by Burns et al.[5] revealed that the two subunits were identical in their amino acid sequence at the N-terminal end for a distance of at least five residues, suggesting that the two subunits may be identical.

Kautsky and Hagerman[19] have studied the 17β-hydroxysteroid of sheep ovary and obtained a 104,000 mol wt for this enzyme by sucrose-gradient centrifugation. There was no indication that the enzyme existed in a monomer-polymer equilibrium although such an equilibrium has been found with the human placental 17β enzyme.[88] Employing gel filtration, Michel et al.[35] obtained a 70,000 mol wt for the sheep ovary 17β-hydroxysteroid dehydrogenase. Polyacrylamide disc gel electrophoresis in the presence of sodium dodecyl sulfate yielded two heavily staining bands with relative mobilities corresponding to molecular weights of 58,000 and 36,000. These authors concluded that this enzyme was a dimer composed of subunits, each having a 36,000 mol wt, and, therefore, very similar to the 17β enzyme of human placenta.

Another soluble 17β-hydroxysteroid dehydrogenase that has been examined is the enzyme of human secretory endometrium. Pollow et al.[92] estimated the molecular weight of this enzyme to be 50 to 54,000 by Sephadex gel filtration. Two values for the molecular weight of the soluble 17β enzyme of human erythrocytes have been reported. Jacobsohn and Hochberg[25] obtained a 75,500 mol wt of this enzyme of Sephadex gel filtration; employing the same technique, Mulder et al.[68] obtained a value of 64,000. The only microsomal 17β-hydroxysteroid dehydrogenase that has been purified and characterized is the pig testis enzyme. A molecular weight of 35,500 was found for this enzyme.[20]

Liu and Kochakian[42] determined the molecular weight of the soluble 17β-hydroxysteroid dehydrogenase of guinea pig kidney to be 35,100, 31,600 and 31,000 by ultracentrifugation, Sephadex gel filtration, and polyacrylamide disc gel electrophoresis, respectively. These results indicated that the guinea pig enzyme lacked subunit structure. Similar molecular weights were obtained for the soluble 17β-hydroxysteroid dehydrogenases of guinea pig and mouse liver. Thaler-Dao et al.[62] have determined the molecular weights of two soluble 17β enzymes of rabbit liver. One had a molecular weight of 35,000 by polyacrylamide disc gel electrophoresis and 34 to 40,000 by Sephadex gel filtration. The second enzyme had a molecular weight of 30,000 by Sephadex gel filtration and yielded two bands of a 25,000 and 30,000 mol wt, respectively, upon sodium dodecyl sulfate-polyacrylamide disc gel electrophoresis. Studies on the corresponding soluble 17α-hydroxysteroid dehydrogenases have been carried out in our laboratory.[87] The molecular weights of the six enzymes purified from rabbit liver cytosol were essentially the same — all values being within 3% of the average value of 39,600. The enzymes possessed no subunit structure. It is interesting to note that the molecular weights of the 17-hydroxysteroid dehydrogenases of the kidney and liver of the guinea pig, rabbit, and mouse are very similar and remarkably close to the molecular weights of the subunits of the human placental and sheep ovarian 17β-hydroxysteroid dehydrogenases.

B. Heterogeneity

Purification of the 17-hydroxysteroid dehydrogenases from tissues of different species has revealed that there are often multiple forms of these enzymes. Isoelectric focusing has proved to be a powerful technique in establishing the microheterogeneity of a number of the enzymes.

Engel and Groman[9] subjected the purified human placental 17β-hydroxysteroid dehydrogenase to isoelectric focusing in polyacrylamide gels and obtained a pattern of three prominent and two faint bands upon staining for either protein or enzyme activity. If isoelectric focusing was carried out in the presence of 8 *M* urea, three protein bands were observed. They concluded that the native enzyme could be formed from three different monomers (present in unequal amounts) interacting to form six dimers.

Pollow et al.[92] have presented evidence indicating a heterogeneity of the cytoplasmic 17β-hydroxysteroid dehydrogenases of normal human endometrium and endometrial carcinoma. Isoelectric focusing of preparations from normal human endometrium in polyacrylamide gels yielded three bands of enzyme activity with either 17β-estradiol or testosterone as substrate. In contrast, the activity pattern from endometrial carcinoma consisted of two, four, or seven bands depending on the substrate (17β-estradiol or testosterone) and cofactor (NAD or NADP) employed. These findings suggested the possibility of different molecular forms of the 17β enzyme in normal and neoplastic endometrium.

Liu and Kochakian[63] carried out polyacrylamide disc gel electrophoresis of the cytosol of guinea pig kidney and observed three strong and two weak bands of 17β-hydroxysteroid dehydrogenase activity. This pattern persisted throughout the purification of the enzyme activity. Storage of the enzyme preparation at −20°C shifted the intensity of the band pattern toward the more acidic forms. This shift could be prevented, or the original pattern restored, by the presence or addition of mercaptoethanol. Electrophoresis at several gel concentrations indicated that the heterogeneity of the 17β-hydroxysteroid dehydrogenase was due to the presence of charge isomers. Kochakian et al.[93] have also reported that the isoenzyme profile of the soluble 17β enzyme of guinea pig kidney was dependent on androgen. The enzyme preparation from the kidneys of normal guinea pigs exhibited three distinct bands following polyacrylamide disc gel electrophoresis and staining for enzyme activity. The two faint bands observed previously[63] were not apparent due to the short incubation period. Analysis of the 17β-hydroxysteroid dehydrogenase activity in animals that had been castrated 35 days previously showed that the intermediate of the three enzyme activity bands had disappeared. When testosterone was administered to the castrated animal, this intermediate band was restored with an intensity similar to that observed with normal animals. The authors concluded that the decrease in 17β enzyme activity and its subsequent restoration after testosterone treatment was due to the disappearance and reappearance of one of the enzyme forms distinguished by polyacrylamide gel electrophoresis.

Both the 17β- and α-hydroxysteroid dehydrogenases of rabbit liver cytosol exhibit heterogeneity. Thaler-Dao et al.[62] separated three 17β-hydroxysteroid dehydrogenase activities which differed in their steroid substrate and pyridine nucleotide specificities. The specificities of these enzymes have been described in an earlier section of this review (Section III. D). Polyacrylamide disc gel electrophoresis of the three partially purified enzyme activities revealed a further heterogeneity in these enzymes. Studies on the 17α enzyme of rabbit liver cytosol carried out in our laboratory revealed the presence of eight forms of this enzyme.[16,17] Five have been purified and their substrate specificities are described in Section IV. The physical properties of these enzymes are very similar; amino acid analysis of each enzyme showed no significant differences in amino acid composition among the enzyme forms.[87] Thus, the observed differences in substrate specificity of these enzyme forms must result from minor amino acid changes in the proteins. It is interesting to note that the primary structure of the subunit of the steroid-active isoenzyme of human liver alcohol dehydrogenase differs from that of the ethanol-active subunit by only six amino acid residues. The amino acid exchanges render the former subunit 3 units more positively charged.[94]

The heterogeneity of the soluble 17α-hydroxysteroid dehydrogenase of rabbit kidney has also been examined in our laboratory for the purpose of comparison with the liver enzyme. This comparison was achieved by subjecting the cytosol of both of these tissues to isoelectric focusing between pH 5 and 7. As is shown in Figure 1, a different pattern of enzyme activity was obtained for the two tissues, indicating that distinct forms of the soluble 17α enzyme are present in rabbit liver and kidney. This isoelectric focusing procedure demonstrated only three enzyme activities in the liver although eight enzyme forms can be resolved by a combination of both DEAE-cellulose chromatography and isoelectric focusing.[17]

C. Mechanism of Action

Several of the properties of the reactions catalyzed by the 17-hydroxysteroid dehydrogenases of mammalian tissues have been documented. Oxidation of the hydroxyl groups at C-17 of the C_{18} and C_{19} steroids generally occurs at an alkaline pH, and the pH optima reported have ranged from a pH of 8.5 for the microsomal 17β-hydroxysteroid dehydrogenase of rat skin[24] to as high as pH 10.4 for this enzyme in dog prostate microsomes.[22] The pH optimum for the reaction in the reductive direction is often significantly lower than that for the oxidation reaction. For example, Langer and Engel[2] studied the soluble 17β enzyme of human placenta and found that the oxidation

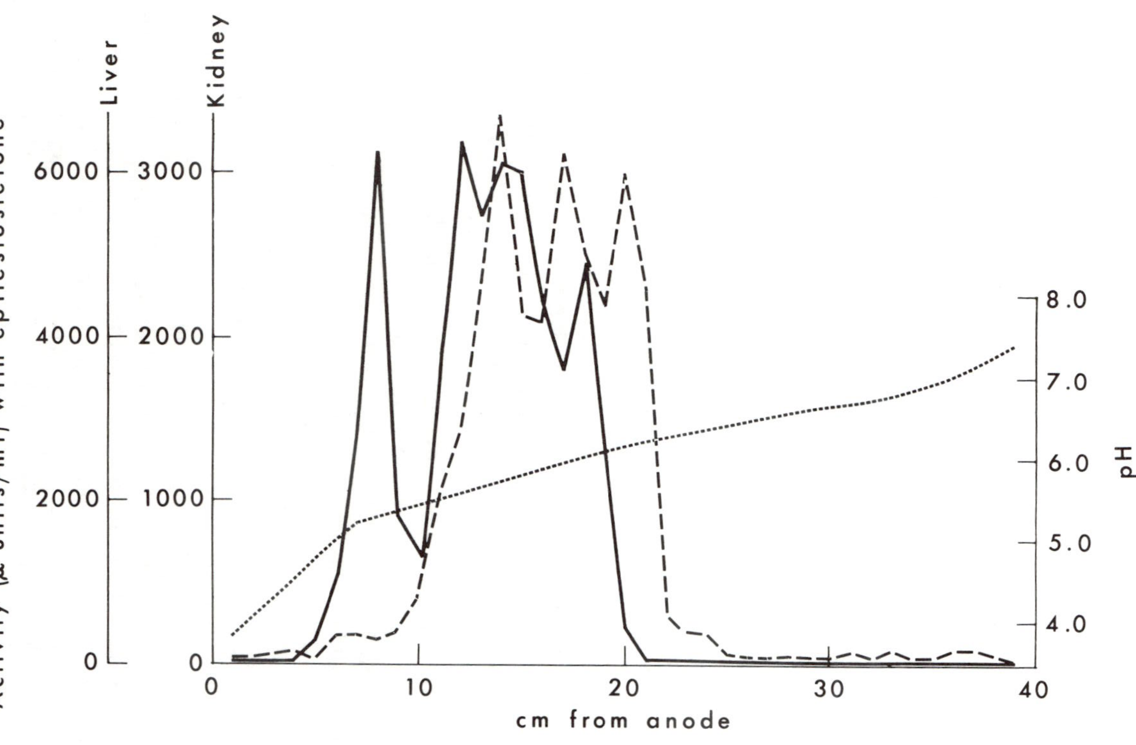

FIGURE 1. Isoelectric focusing of the soluble 17α-hydroxysteroid dehydrogenases of rabbit liver and kidney. Enzyme activity in the liver (---) and kidney (—), pH (· · · ·).

of 17β-estradiol by this enzyme occurred optimally at pH 10, while the pH optimum for estrone reduction was 6.2. In some cases, enzymes having a dual nucleotide specificity display minor differences in pH optima with NAD and NADP.[22] Jacobsohn and Hochberg,[25] in studying the 17β-hydroxysteroid dehydrogenase of human erythrocytes, observed a pH optimum of 9.7 with 17β-estradiol and 9.0 with 17β-estradiol 3-sulfate. With microsomal preparations from pig testis, the pH optimum of the 17β has been shown to shift to a more alkaline pH as the incubation temperature was increased.[20]

The kinetic parameters for the interaction of steroid substrates and pyridine nucleotides with a number of 17-hydroxysteroid dehydrogenases have been determined; in some cases, this information has been used to establish the molecular mechanism of the enzyme-catalyzed reaction. Warren and Crist[95] have carried out a kinetic analysis of the site specificity of the soluble 17β-hydroxysteroid dehydrogenase of human placenta in terms of both cofactor (NAD and NADP) and steroid substrate (17β-estradiol and -estradiol 3-sulfate). Their data, based on initial velocity studies, indicated that both steroids were bound at a single active site. NAD and NADP had an active site in common, but a second NAD binding site was also present. They postulated an ordered mechanism in which either steroid or cofactor bound first to the enzyme; supportive evidence favored the binding of steroid first. The steroid and cofactor formed a ternary complex with the enzyme before any catalytic step occurred. In more recent studies, Betz[96] has determined the reaction mechanism of the human placental 17β-hydroxysteroid dehydrogenase by the method of equilibrium rate exchange and found that either the steroid or nucleotide formed binary complexes with the enzyme. This data was consistent with a random addition of substrates to the enzyme in contrast to an ordered mechanism in which the binding of one substrate was prerequisite for the binding of the second.

Michel et al.[35] carried out a kinetic analysis of the reaction catalyzed by the 17β enzyme of sheep ovary which indicated a compulsory-order mechanism with the pyridine nucleotide binding first. This mechanism was supported by the observation that the binding of the coenzyme enhanced the affinity of the enzyme for the steroid substrates by 2- to 30-fold.

Further studies on the mechanism of the 17-hydroxysteroid dehydrogenase reactions have involved an examination of the stereospecificity of hydride transfer from the pyridine nucleotide coenzymes to steroid substrates. Jarabak and Talalay[97] found that the soluble 17β-hydroxysteroid dehydrogenase of human placenta catalyzed the transfer of the 4-pro-S hydrogen of either NADH or NADPH to the steroid substrate. This stereochemistry was also observed with the microsomal 17β enzyme of pig testis by Inano and Tamaoki.[57] Achtar et al.[98] summarized the stereospecificity of hydrogen transfer of a number of pyridine nucleotide-dependent mammalian dehydrogenases and reductases. These authors stated that hydrogen transferred to the α-face of the steroid molecule orginated from the 4-pro-S hydrogen of the pyridine nucleotide, while transfer to the β-face was from the 4-pro-R hydrogen of the pyridine nucleotide. However, exceptions to this general statement have been noted. George et al.[99] found that both the 17β- and α-hydroxysteroid dehydrogenases of chicken liver catalyzed the transfer of hydrogen from the estradiol epimers to the 4-pro-S position of NADP. In another study, Groman et al.[100] demonstrated that, in the reduction of 17β-hydroxy-5α-androstan-3-one to the 3β-alcohol catalyzed by horse liver alcohol dehydrogenase, the 4-pro-R hydrogen of NADH was transferred.

D. Studies on the Active Site of the Soluble 17β-Hydroxysteroid Dehydrogenase of Human Placenta

In recent years, studies concerning the active site of the soluble 17β-hydroxysteroid dehydrogenase of human placenta have been carried out by affinity labeling with steroid and pyridine nucleotide analogues having reactive groups that can form covalent bonds with amino acid residues in the enzyme protein.

Biellman et al.[101] investigated the active site of the human placental 17β enzyme with a NAD analogue, 3-chloroacetylpyridine-adenine dinucleotide. They found that, in addition to functioning as a hydride acceptor, this analogue inactivated and alkylated the enzyme. The data indicated that the alkylation of a cysteine residue occurred with the stoichiometry of 1 mol/mol of enzyme subunit, leading them to suggest that this cysteine residue was probably part of the active site of the enzyme and was located in the proximity of the pyridine ring of the coenzyme. This study supported the earlier observations of Nicolas and Harris,[102] who had studied the inactivation of the placental 17β-hydroxysteriod dehydrogenase with ^{14}C- labeled *N*-ethylmaleimide. They determined that this inhibitor reacted with at least two cysteine residues out of a total of twelve per mole of enzyme. Inactivation with *N*-ethylmaleimide was prevented by the addition of NADP but not by 17β-estradiol, thus suggesting that the cysteines comprised part of the coenzyme binding site of the enzyme.

Pons et al.[103] have employed derivatives of estrone and 17β-estradiol as affinity labels. They observed that, in the presence of NADPH, 3-iodoacetoxyestrone produced a specific alkylation of a single histidine residue in the soluble 17β-hydroxysteroid dehydrogenase of human placenta. The labeling was less specific in the absence of the reduced coenzyme and, in addition to the histidine residue, two cysteine residues were also alkylated. Recently, Pons et al.[104] have carried out affinity labeling of the placental enzyme using derivatives of 17β-estradiol which have alkylating groups at nine different positions on the steroid nucleus. When the alkylating side chain was situated at C-3 of the steroid, a rapid alkylation of a histidine residue in the enzyme occurred. This alkylation dramatically decreased when the alkylating side chain was shifted toward rings B or D of the substrate analogue. These data indicated that the histidine residue was probably located in the vicinity of ring A of the steroid substrate which was bound at the active site of the enzyme. A cysteine residue located at the active site was alkylated more readily when the alkylating side chain was at C-16 than when it was at C-3. In light of the data obtained by Biellman et al.[101] with 3-chloroacetylpyridine-adenine dinucleotide, this result suggests that this cysteine residue is located in the catalytic region of the active site at the junction of ring D of the steroid nucleus with the nicotinamide moiety of the coenzyme.

Groman et al.[105] have carried out affinity labeling of the human placental 17β-hydroxysteroid dehydrogenase with 3-bromoacetoxyestrone. In these studies, direct evidence that the affinity label was covalently bound to the enzyme at the active site was obtained by showing that the bound steroid was reduced to 17β-estradiol by the enzyme. These authors used the term "catalytic competence" to describe the ability of an enzyme to catalyze its normal reaction with a covalently bound substrate and stated that this ability was an important criterion for establishing that the affinity labeling occurred at the substrate-binding site and not at noncatalytic steroid-binding site.

VI. REGULATION OF THE 17-HYDROXYSTEROID DEHYDROGENASES

A. Development and Sexual Differentiation

The 17β-hydroxysteroid dehydrogenase activity in some tissues of mammalian species has been shown to be dependent on age. For example, Bongiovanni and Marino[106] examined this enzyme activity in the liver of rabbits of various ages. While no 17β enzyme activity was detected in fetal liver (30 to 33 days of gestation), it was present in the liver of young animals (20 days old) and mature animals (17 to 19 months old). In some species and tissues, the age-dependent development of 17β-hydroxysteroid dehydrogenase activity is accompanied by a sex-specific differentiation of this enzyme activity. Ghraf et al.[107] have measured 17β enzyme activity in the kidney, liver, adrenal, and gonads of rats 15 to 120 days old. Activity was measured in both the microsomal and soluble tissue fractions. In all of the tissues studied, the enzyme activity of at least one of the subcellular fractions displayed an age-dependent development, reaching significantly higher levels in 60-day-old animals. The only exceptions noted were cytosol of the adrenal, the microsomes of the kidney, and the cytosol of female liver, in which no increases in enzyme activity were observed. There was also a sexual differentiation of enzyme activity in most tissues exhibiting an age-dependent enzyme development. This differentiation occurred between 30 and 60 days after birth. 17β-Hydroxysteroid dehydrogenase activity in the kidney, adrenal, and gonads was significantly higher in female rats than in male rats whereas the enzyme activity in the cytosol of male liver was higher than that observed in the cytosol of female liver.

A relationship between the 17β-hydroxysteroid dehydrogenase activity of rat submaxillary gland and the age and sex of the animal has been reported by Baldi and Charreau.[108] The conversion rate of testosterone to androstenedione was higher in the female gland than in the male, a difference which was not observed when 17β-estradiol was the substrate. Androstenedione production from testosterone decreased gradually with increasing age of the animal and was maximal in the rat fetus. These authors suggested that the sex differences in 17β-hydroxysteroid dehydrogenase activity were related to the histological and functional differences in the tubular ducts of the submaxillary glands. These ducts were taller and more columnar in adult male rats than in females. Since these histological changes are caused by testosterone, the diminished oxidation of this androgen in the submaxillary gland of the male rat (when compared to the female) would lead to a greater effect of testosterone on the male gland.

B. Hormonal Regulation

The effects of castration and/or hormone administration on the activities of enzymes involved in steroid metabolism in animals at various stages of development have been examined by a number of investigators.[109-112] Schriefers et al.[109] observed a sexual differentiation in the soluble 17β-hydroxysteroid dehydrogenase of rat liver — activity was significantly higher in the male than in the female rat. This difference was abolished by the administration of estradiol benzoate to 1-day-old male rats, which caused the liver enzyme activity to decrease to female levels. Subsequent treatment of the neonatally estrogenized male rats with testosterone restored the normal difference in enzyme activity. The data also indicated that only a neonatal androgen stimulus of short duration was required to produce a sexual differentiation of the soluble 17β-hydroxysteroid dehydrogenase of rat liver. Castration of 21-day-old male rats did not significantly decrease this enzyme activity. Einarsson et al.[111] have divided the enzymes of steroid metabolism in liver tissue into three groups based on the mechanisms which regulate their activities. The 17β-hydroxysteroid dehydrogenase of rat liver belongs to the group of enzymes that are irreversibly "imprinted" or "programmed" by androgens during the prepubertal period and reversibly stimulated by androgens postpubertally. Hormone secretion by the testis at some time between birth and puberty causes an increase in the activity of this enzyme in the liver of the male rat when compared to the female. Neonatal castration of male rats abolishes the sex difference in enzyme activity while postpubertal castration reduces, but does not abolish, this difference. The decrease in 17β enzyme activity observed in the latter situation can be reversed by testosterone administration.

The 17α-hydroxysteroid dehydrogenase of rat liver is also sex specific and activity toward androstenedione is higher in male than in female rats.[112] This higher level of activity in male rats is dependent on androgens and irreversibly induced at birth by androgens. This androgen dependency could explain the earlier observation of Kraulis and Clayton[113] that 5α-androstanedione-3β, 17α-diol 3-sulfate was a metabolite of testosterone in the blood of male rats. It was produced in significantly greater amounts in adult male rats that had been castrated 2 weeks after birth than in those that had been castrated on the first postnatal day. Androgenic induction of hepatic 17α-hydroxysteroid dehydrogenase activity may also be of significance in the human since a high excretion of 17α-hydroxy-C_{19} steroids has been observed in association with virilized states in women.[33,114]

Lax et al.[115] and Ghraf et al.[116] have examined the role of the gonads and hypophysis in the regulation of the soluble 17β-hydroxysteroid dehydrogenase of rat liver. Activity was significantly higher in intact male rats than in female rats and was not decreased in the male by castration. Hypophysectomy or administration of 17β-estradiol to male rats decreased the enzyme activity to the female level, while gonadectomy or testosterone administration to female rats increased the soluble hepatic 17β-hydroxysteroid dehydrogenase activity to the male level. These authors termed the enzyme in rat liver cytosol as estrogen dependent, since gonadectomy of male animals had no effect on the enzyme activity whereas gonadectomy of female animals resulted in a masculinization of the enzyme activity. Ghraf et al.[117] conducted similar studies on the soluble 17β enzyme of rat kidney and concluded that this activity was androgen dependent. In intact animals, enzyme activity was higher in females than in males. Castration of the male rat increased the activity to the female level, an effect reversible by administration of testosterone. The administration of testosterone to either intact or gonadectomized female animals decreased the enzyme activity to the male level. In male and female rats, hypophysectomy resulted in a decrease in the 17β-hydroxysteroid dehydrogenase activity of the kidney to levels far below that observed with intact animals and no sex difference in the activity was observed.

In other studies on the hormonal regulation of 17β enzyme activity, Musto et al.[118] observed that the activity of this enzyme in the mouse testis was increased by prolactin. The sterile, hereditary dwarf, male mouse which is deficient in prolactin while having normal level of LH and FSH was used in this study. Implantation of a normal mouse pituitary in these animals increased the 17β-hydroxysteroid dehydrogenase activity in the testis to a value 244% that of control animals, while subcutaneous injection of prolactin increased enzyme activity to 186% that of control animals. Since the onset of puberty in male animals is accompanied by an increase in 17β enzyme activity, prolactin may be involved in this sexual maturation.

17β-Hydroxysteroid dehydrogenase activity is absent or very low in the uterus of the immature rat[119] or rabbit.[120] However, this enzyme activity has been observed in incubations of uterine preparations of adult rats and rabbits.[120-122] Studies on the 17β enzyme activity of the immature rabbit and the ovariectomized adult rabbit have shown that the uterine enzyme is stimulated by 17β-estradiol. In the female rat, the oxidation of 17β-estradiol is estrous cycle-dependent and decreases dramatically upon ovariectomy.[121] In studies on the conversion of 17β-estradiol to estrone by rat uterine tissue slices, Wenzel et al.[54] found that maximal oxidation of estrogen occurred at the first day of estrus and minimally during proestrus. They suggested that the increased oxidation during estrus resulted from an increased secretion of estrogen during proestrus, which in turn caused an increase in the synthesis of the uterine 17β-hydroxysteroid dehydrogenase.

C. Nonhormonal Regulation

Many of the 17-hydroxysteroid dehydrogenases exhibit broad substrate specificities and act on a number of structurally related C_{18} and C_{19} steroids. The consequence of this broad substrate specificity is that one substrate can inhibit, usually in a competitive manner, the oxidation or reduction of another substrate. For example, Oshima et al.[123] showed that, in cell-free homogenates of human testis, 17β-hydroxysteroid dehydrogenase activity toward dehydroepiandrosterone was inhibited by androstenedione, which is itself a substrate for this enzyme. Studies have also shown that 17β enzymes may be inhibited by nonsubstrate steroids. The soluble 17β-hydroxysteroid dehydrogenase of human placenta is specific for 17β-hydroxysteroids having an aromatic ring A. However, the oxidation of 17β-estradiol by this enzyme can be competitively inhibited by 17α-estradiol, estriol, and diethylstilbestrol — steroids that are not substrates for this enzyme.[45,89] The oxidation of 17β-estradiol by the 17β enzyme of rat skin is inhibited by a number of related C_{18} and C_{19} steroids possessing a 17β-hydroxyl group.[24] The inhibition of 17β-hydroxysteroid dehydrogenases by steroids that may or may not be substrates often occurs with physiological concentrations of these steroids. Although inhibition has only been demonstrated in in vitro studies, it is possible that these effects may have in vivo significance. The oxidoreduction rate of the functional group at C-17 of a steroid may be modulated by the presence of other steroids which may interact with the 17β enzyme.

Inhibition of 17-hydroxysteroid dehydrogenases by nonsteroid biological compounds has also been demonstrated. Sulimovici and Lunenfeld[124] found that the NAD-dependent oxidation of testosterone by the microsomal 17β-hydroxysteroid dehydrogenase of rat testis was inhibited by 3′,5′-cyclic adenosine monophosphate (c-AMP). The NADP-linked oxidation of testosterone was not inhibited by c-AMP, nor was the reduction of androstenedione in the presence of NADPH. It is possible that the effect of c-AMP on this NAD-dependent enzyme is related to the regulation of steroidogenesis in this tissue by the pituitary hormones FSH and LH, which have been shown to increase the c-AMP levels in the testis.[125,126]

Inhibition of the soluble 17β-hydroxysteroid dehydrogenase of human placenta by ATP has recently been reported by Shaw and Jeffery.[127] ATP was a much stronger inhibitor of estrone or 16α-hydroxyestrone reduction when NADH rather than NADPH was employed as the coenzyme. Inhibition occurred at physiological concentrations of ATP. These results suggest that there may be a relationship between energy metabolism in the placenta and oxidoreduction at C-17 of the estrogens.

VII. BIOLOGICAL ROLE OF THE 17-HYDROXYSTEROID DEHYDROGENASES

A. Role in Target Tissues

Estrogens and androgens possessing a 17β-hydroxyl group are biologically the most potent steroids with regards to their effects on the growth and function of target tissues. For example, in the rabbit, 17β-estradiol is ten times more active than 17α-estradiol in promoting uterine growth.[128] Therefore, the 17-hydroxysteroid dehydrogenases present in hormone-responsive tissues could help to regulate the biological action of estrogens or androgens in these tissues by structurally altering the steroid molecules at C-17. Baird et al.[129] have used the term prehormone to describe compounds usually secreted by endocrine glands which have little or no biological potency but are converted to more active compounds in peripheral tissues, including the target organs.

A number of studies on the interconversion of estrogens in the uteri of different mammalian species have been undertaken. Russell and Thomas[130] employed explants of rabbit uterus, vagina, and skeletal muscle to determine the uptake and metabolism of estrone and 17β-estradiol. When tissue from the uterus or vagina was cultured in the presence of either estrone or 17β-estradiol, the tissue contained predominantly 17β-estradiol. In contrast, skeletal muscle showed only a limited interconversion of the estrogens. 17α-Hydroxysteroid dehydrogenase activity was extremely low in the uterus, a result consistent with the low biological activity of 17α-estradiol in this tissue.[128]

Similar results were obtained by Crocker and Thomas[131] in studies on the uptake and interconversion of estrogens in the rabbit in vivo. If either 17β-estradiol or estrone was infused into the animal, the uterus and vagina contained predominantly 17β-estradiol. The 17β-estradiol: estrone ratios of the pituitary and other tissues approximated that of the plasma. These observations have been confirmed by Tseng et al;[132] who studied the interconversion of estrone and 17β-estradiol in rabbit uterus by both *in situ* and in vitro perfusion. The rate constant for the conversion of estrone to 17β-estradiol was 10 to 20 times larger than the rate constant for the reverse reaction. Studies involving the rabbit indicate that the 17β-hydroxysteroid dehydrogenase of the uterus may function in maintaining the active form of the estrogen, namely, 17β-estradiol, in this tissue. Thus, any regulation of this enzyme activity could affect the biological action of the estrogens on the uterus. A similar role could be ascribed to the 17β-hydroxysteroid dehydrogenases of other tissues responsive to either estrogens or androgens. This function is important since it is known that the mechanism of action of androgens and estrogens involves the intial binding of the active hormone to a receptor protein with a high affinity for the 17β-hydroxysteroid.[133]

B. Role in Other Tissues

As discussed in Section III, the presence of 17-hydroxysteroid dehydrogenase activity in tissues other than endocrine or steroid-responsive tissues has been well documented. This is particularly true of the 17β enzymes which have a ubiquitous tissue distribution in many mammalian species. While a role for the 17-hydroxysteroid dehydrogenases in endocrine and steroid-responsive tissues can be readily ascribed, a function for these enzymes in other tissues is more difficult to establish. Studies have shown that many tissues may respond to estrogen or androgen administration even though these tissues are not considered target tissues in the classical sense. Estrogen induction of the synthesis of egg yolk proteins (in particular, phosvitin) in the liver of laying hens, immature chicks, and roosters has been reported.[134,135] A cytoplasmic, 17β-estradiol-binding protein has been identified in chicken liver cytosol.[136,137] Van Beurden-Lamers et al.[138] have detected high-affinity 17β-estradiol-binding macromolecules in the cytoplasm of the liver, adrenal, pituitary, prostate, epididymis, and testis interstitial tissue of the male rat. These receptors had a low affinity for estrone or estriol. Thus, one function of the 17β-hydroxysteroid dehydrogenases in these tissues may be to convert estrone to 17β-estradiol prior to the binding of the hormone to the cytosol receptor.

The 17β enzymes present in various tissues also play a role in the inactivation of estrogens and androgens. Oxidation of the 17β-hydroxyl group of these steroid hormones dramatically decreases their biological potency. In some species, further inactivation of estrogens and androgens is achieved by reduction of 17-ketosteroid to 17α-hydroxysteroid, e.g., in ruminants and the rabbit, where 17α-estradiol is the major urinary metabolite of 17β-estradiol.[10] Thus, the primary role of the 17α enzyme in these species is in the catabolism of estrogens and androgens. The liver and kidney are the major tissues involved in this metabolic pathway.

Studies in our laboratory indicate that, in the rabbit liver, there may be a special relationship between oxidoreduction at C-17 of the estrogen molecule and conjugation with glucuronic acid at C-3. In rabbit liver cytosol, 17-hydroxysteroid dehydrogenase activity favors interconversion of estrone and 17β-estradiol rather than formation of 17α-estradiol. However, with the glucuronide derivatives of the steroids, formation of the 17α-epimer predominates (Table 4). Rabbit liver cytosol contains a 17α-hydroxysteroid dehydrogenase that, with estrogen substrates, is specific for the 3-glucuronide derivative.[17,87] These findings may have significance with respect to the control of estrogen metabolism in the rabbit. This metabolic pathway involves formation of 17α-estradiol and conjugation first with glucuronic acid at C-3 and then with *N*-acetylglucosamine at C-17.[11,139,140] The transfer of *N*-acetylglucosamine to the steroid molecule is very specific and requires prior formation of the 17α-epimer and prior conjugation at C-3 with glucuronic acid. The double conjugate formed is subsequently excreted in the urine. Therefore, in rabbit liver, conjugation of the estrogen molecule at C-3 with glucuronic acid directs metabolism towards elimination of this steroid owing to the presence of a 17α-hydroxysteroid dehydrogenase specific to the steroid glucuronide derivative.

The relationship between steroid conjugation and oxidoreduction demonstrated in the rabbit may function in other pathways of steroid metabolism in different mammalian species. Both estrogens and androgens conjugated at C-3 with either sulfuric or glucuronic acid have been shown to undergo oxidoreduction at C-17.[141,142] In studies on the soluble 17β-hydroxysteroid dehydrogenase of guinea pig liver, Milgrom and Baulieu[141] found that the C_{19} steroid conjugates androstenediol 3-sulfate and 3-glucuronide were "directly" oxidized to the corresponding 17-ketosteroids. Moreover, dehydroepiandrosterone 3-sulfate or 3-glucuronide also inhibited the oxidation of testosterone, which suggests that the steroid conjugates could affect or regulate testosterone metabolism by interaction in vivo with the 17β-hydroxysteroid dehydrogenase. However, the relationship between steroid conjugation and oxidoreduction is not limited to the 17-hydroxysteroid dehydrogenases. Robel et al.[143] compared the urinary metabolites obtained after administration of testosterone and testosterone glucuronide to human subjects and found that those of administered testosterone were 5α-metabolites while testosterone glucuronide formed 5β-metabolites. Thus, the presence of a glucuronic acid moiety at C-17 of testosterone affected ring A reduction of this androgen.

C. Transhydrogenase Activity

Villee and Hagerman[144] first observed the 17β-estradiol stimulation of the enzymatic transfer of hydrogen from NADPH to NAD in human placental extracts. Talalay and Williams-Ashman[145] showed this to be a transhydrogenation. Further studies[146,147] indicated that the soluble 17β-hydroxysteroid dehydrogenase was responsible for the transhydrogenase activity in human placenta due to the enzyme's dual pyridine nucleotide specificity. 17β-Estradiol acted as a coenzyme, and its C-17 hydroxyl group was cyclically oxidized and reduced during the transhydrogenation reaction. These researchers concluded that the 17β enzyme activity was responsible for most of the 17β-estradiol-mediated transfer of hydrogen between the pyridine nucleotides in the cytosol of human placenta.[147] However, Hagerman and Villee[148] demonstrated that an estrogen-sensitive transhydrogenase was also present in human placenta which did not catalyze the reduction of 17β-estradiol. They stated that this enzyme accounted for nearly half of the observed pyridine nucleotide transhydrogenation in crude placental extracts.[149] The presence of this distinct transhydrogenase was confirmed by Karavolas et al.,[150] who separated this enzyme from the 17β-hydroxysteroid dehydrogenase and found different kinetic and physical properties. Estrogen-dependent, pyridine nucleotide transhydrogenation has also been demonstrated in other tissues of the human and other mammalian species.[151]

Hagerman[88] has shown that the purified, soluble 17β -hydroxysteroid dehydrogenase of human placenta can exist in multiple polymeric forms, (i.e., as a monomer, dimer, and trimer). The enzyme concentration was the only factor found to affect the relative concentrations of these enzyme forms. While all enzyme forms exhibited transhydrogenase activity, only the monomer and dimer displayed dehydrogenase activity. Consequently, Hagerman[88] suggested (1) the dehydrogenase and transhydrogenase activities of the human placental 17β-hydroxysteroid dehydrogenase were directly dependent on the intracellular concentration of the enzyme and (2) the metabolic function of the enzyme could be regulated by altering its local intracellular concentration.

Both the transhydrogenase activity of the 17β-hydroxysteroid dehydrogenase and the 17β-estradiol-sensitive transhydrogenase could indirectly regulate estrogen biosynthesis in the placenta by controlling the intracellular concentration of NADPH. The aromatization of androstenedione requires 3 mol of NADPH per mole of estrogen formed.[152]

The transhydrogenase activity of other 17-hydroxysteroid dehydrogenases exhibiting a degree of dual pyridine nucleotide specificity has been examined. Inano and Tamaoki[57] reported that the purified microsomal 17β enzyme of pig testis catalyzed no detectable transhydrogenation from either NADPH to NAD or NADH to NADP. This lack of activity was probably due to the high affinity of the enzyme for NADP(H) compared to NAD(H). Kautsky and Hagerman[19] found that the substrate-mediated transhydrogenase activity of the 17β enzyme of sheep ovary was only 6% that of the dehydrogenase activity. This latter activity was about one third the activity obtained with the analogous human placental enzyme assayed under the same conditions.

A role for the 17-hydroxysteroid dehydrogenases in the transfer of hydrogen between steroid molecules has also been described. In studies involving the oxidation of 17β-estradiol by rat uterus, Wenzel et al.[54] observed an increased rate of oxidation of this estrogen when 5α-androst-3-en-17-one was added to the incubation. The rate increase was due to the specific enzymatic transfer of hydrogen from 17β-estradiol to 5α-androst-3-en-17-one to produce 5α-androst-3-en-17-ol. A similar transfer of hydrogen was observed with rat liver and kidney preparations.[153] Pollow et al.[154] have shown that the 17β- and 20α-hydroxysteroid dehydrogenases of human placenta mitochondria could, acting together, catalyze the transfer of hydrogen between 17β-estradiol and progesterone. They suggested that, due to this specific hydrogen transfer, the steroids could act as hydrogen-carrying coenzymes by forming part of a chain of hydrogen carriers. Thus, they could function in the transfer of hydrogen between the cytoplasm and mitochondria of the placental cell.

D. Role in Steroid Transport

The initial step in the mechanism of steroid hormone action is the binding of the steroid to a specific cytoplasmic receptor protein present in the target cell.[155,156] It is usually assumed that the steroid enters the cell by passive diffusion through the cell membrane. However, Baulieu[155] has reported that, in some instances, there may be a protein-facilitated diffusion of the steroid through the plasma membrane. Moreover, plasma membrane protein receptors for 17β-estradiol have been demonstrated recently in the rat uterus and liver.[157] Watanabe and co-workers[158] have been conducting studies on steroid transport through membranes and employing (as a model system) membrane vesicles obtained from the bacterium *Pseudomonas testosteroni*. This organism can grow on C_{19} and C_{21} steroids and, during adaptive growth, there is an induction of steroid dehydrogenase activities.[159] Watanabe and Po[160] demonstrated that the membrane-bound 3(17)β-hydroxysteroid dehydrogenase of *P. testosteroni* functioned in steroid transport through membrane vesicles of this bacteria. Steroid transport was dependent on NAD, and this pyridine nucleotide was reduced to NADH during the process. Transport of testosterone or dihydrotestosterone resulted in their conversion to androstenedione and androstanedione, respectively; no intravesicular testosterone or dihydrotesterone could be detected. Employing sulfhydryl and disulfide reagents, Lefebvre et al.[161] distinguished two components of the steroid transport system in *P. testosteroni* membrane vesicles: a 17β-hydroxysteroid dehydrogenase and a transport protein. Further studies demonstrated that the electron transport chain was also involved in this process.[162] In the uptake of steroid by the membrane vesicles, the 17β enzymes and the electron transport chain could be linked through the production and utilization of NADH, with the steroid functioning by its oxidation as the physiological electron donor via NADH.

This chapter has reviewed many mammalian 17-hydroxysteroid dehydrogenases that are components of the membrane structure of cellular organelles. It is tempting to speculate that these enzymes may be involved in the intracellular transport of steroids.

VIII. SUMMARY

Studies on the chemical and physical properties of the mammalian 17-hydroxysteroid dehydrogenases have been limited primarily due to the difficulty in obtaining sufficient quantities of these enzymes to carry out such studies. Purification is difficult since these enzymes represent a very small percentage of the total protein in a tissue and, in some cases, the enzymes are associated with the membrane structures of the cells. Of the 17β-hydroxysteroid dehydrogenases, the soluble enzyme from human placenta has been successfully purified in quantities permitting studies on its physical and chemical properties. The structural features of the steroid substrates required for binding to the enzyme are now known, and information on the structure of the active site is being obtained by affinity labeling. In addition to delineating the mechanism of the reactions catalyzed by the 17-hydroxysteroid dehydrogenases, studies such as those being carried out with the enzyme from human placenta may provide an insight into the general features of steroid-protein interaction. This information is needed to establish how steroid-protein interaction is related to the biological action of these hormones.

Although the 17-hydroxysteroid dehydrogenases have many properties in common, there are differences in subcellular distribution, substrate specificity, and pyridine nucleotide specificity among the enzymes of different tissues which probably reflect the specific functions of the enzymes in various tissues. Therefore, detailed studies on the properties of individual 17-hydroxysteroid dehydrogenases could provide information on the role of these enzymes in mammalian tissues and on the mechanisms whereby the activities of these enzymes are regulated. Although the structural alterations of the steroid molecule (which are caused by the action of the 17-hydroxysteroid dehydrogenases) are relatively subtle, these changes greatly effect the biological activity of the steroid hormones. Therefore, the 17-hydroxysteroid dehydrogenases play a key role in both the mechanism of action and the metabolism of steroid hormones.

ACKNOWLEDGMENT

I wish to thank Mr. Peter Lau and Mrs. Irena Brglez for their valuable contribution to the studies on the 17-hydroxysteroid dehydrogenases of rabbit tissues that were carried out in my laboratory. This research was supported by a grant (MT-4494) to the author from the Medical Research Council of Canada.

REFERENCES

1. **Tamaoki, B.,** Steroidogenesis and cell structure, biochemical pursuit of sites of steroid biosynthesis, *J. Steroid Biochem.,* 4, 89, 1973.
2. **Langer, L. J. and Engel, L. L.,** Human placental estradiol-17β dehydrogenase; concentration, characterization and assay, *J. Biol. Chem.,* 233, 583, 1958.
3. **Jarabak, J.,** Soluble 17β-hydroxysteroid dehydrogenase of human placenta, in *Methods in Enzymology,* Vol. 15, Clayton, R. B., Ed., Academic Press, New York, 1969, 746.
4. **Karavolas, H. J., Baedecker, M. L., and Engel, L. L.,** Human placental 17β-estradiol dehydrogenase, purification and partial characterization of the diphosphopyridine nucleotide(triphosphopyridine nucleotide)-linked enzyme, *J. Biol. Chem.,* 245, 4948, 1970.

5. **Burns, D. J. W., Engel, L. L., and Bethune, J. L.,** Amino acid composition and subunit structure. Human placental 17β-estradiol dehydrogenase, *Biochemistry,* 11, 2699, 1972.
6. **Nicolas, J. C., Pons, M., Descomps, B., and Crastes de Paulet, A.,** Affinity chromatography: large-scale purification of the soluble estradiol-17β dehydrogenase of human placenta, *FEBS Lett.,* 23, 175, 1972.
7. **Chin, C. C. and Warren, J. C.,** Purification of estradiol-17β dehydrogenase from human placenta by affinity chromatography, *Steroids,* 22, 373, 1972.
8. **Chin, C. C., Dence, J. B., and Warren, J. C.,** Crystallization of human placental estradiol 17β-dehydrogenase, *J. Biol. Chem.,* 251, 3700, 1976.
9. **Engel, L. L. and Groman, E. V.,** Human placental 17β-estradiol dehydrogenase: characterization and structural studies, in *Recent Progress in Hormone Research,* Vol. 30, Greep, R. O., Ed., Academic Press, New York, 1974, 139.
10. **Velle, W.,** Metabolism of estrogenic hormones in domestic animals, *Gen. Comp. Endocrinol.,* 3, 621, 1963.
11. **Collins, D. C., Williams, K. I. H., and Layne, D. S.,** Further studies on the nature of phenolic steroids in the rabbit and their mode of conjugation, *Arch. Biochem. Biophys.,* 121, 609, 1967.
12. **Breuer, H. and Pangels, G.,** Stoffwechsel von Oestron, Oestradiol-17α und Oestradiol-17β in der Kaninchenleber, *Acta Endocrinol.,* 33, 532, 1960.
13. **Williams, K. I. H., Henry, D. H., Collins, D. C., and Layne, D. S.,** Metabolism of 4-^{14}C-17β-^{3}H-estradiol-17α and 16-^{14}C-17α-^{3}H-estradiol-17β by the rabbit, *Endocrinology,* 83, 113, 1968.
14. **Szamatowicz, M., Drosdowsky, M., and Jayle, M. F.,** The role of testosterone and androstenedione as precursors of epitestosterone in guinea pigs, *Acta Endocrinol.,* 67, 187, 1971.
15. **Ball, P. and Breuer, H.,** Purification and characterization of a cytoplasmic 17-hydroxysteroid: NAD(P) oxidoreductase from rabbit liver, *Hoppe-Seylers Z. Physiol. Chem.,* 351, 1011, 1970.
16. **Hasnain, S. and Williamson, D. G.,** The separation and partial purification of the soluble 17α- and 17β-hydroxysteroid dehydrogenases of rabbit liver, *Can. J. Biochem.,* 52, 120, 1974.
17. **Hasnain, S. and Williamson, D. G.,** Purification of multiple forms of the soluble 17α-hydroxysteroid dehydrogenase of rabbit liver, *Biochem. J.,* 147, 457, 1975.
18. **Renwick, A. G. C. and Engel, L. L.,** The partial purification of 17α- and 17β-estradiol dehydrogenase activities from chicken liver, *Biochim. Biophys. Acta,* 146, 336, 1967.
19. **Kautsky, M. P. and Hagerman, D. D.,** 17β-Estradiol dehydrogenase of ovine ovaries, *J. Biol. Chem.,* 245, 1978, 1970.
20. **Inano, H. and Tamaoki, B.,** Purification and properties of NADP-dependent 17β-hydroxysteroid dehydrogenase solubilized from porcine — testicular microsomal fraction, *Eur. J. Biochem.,* 44, 13, 1974.
21. **Ryan, K. J. and Engel, L. L.,** The interconversion of estrone and estradiol by human tissue slices, *Endocrinology,* 52, 287, 1953.
22. **Hussein, K. A. and Kochakian, C. D.,** 17β-Hydroxy-C_{19}-steroid dehydrogenase activity of dog prostate, solubilization and partial purification and characterization, *Steroids,* 12, 589, 1968.
23. **Aoshima, Y. and Kochakian, C. D.,** Activity, intracellular distribution and some properties of 17β-hydroxy-C_{19}-steroid dehydrogenases in liver and kidney, *Endocrinology,* 72, 106, 1963.
24. **Davis, B. P., Rampini, E., and Hsia, S. L.,** 17β-Hydroxysteroid dehydrogenase of rat skin, substrate specificity and inhibitors, *J. Biol. Chem.,* 247, 1407, 1972.
25. **Jacobsohn, G. M. and Hochberg, R. B.,** 17β-Hydroxysteroid dehydrogenase from human red blood cells, *J. Biol. Chem.,* 243, 2985, 1968.
26. **Delin, S. and Porath, J.,** Purification of α- and β-hydroxysteroid dehydrogenase from *Pseudomonas testosteroni* by gel filtration, *Biochim. Biophys. Acta,* 67, 197, 1964.
27. **Markert, C. and Trager, L.,** 17β-Hydroxysteroid dehydrogenase activity in *Streptomyces hydrogenans, Acta Microbiol. Acad. Sci. Hung.,* 22, 503, 1975.
28. **Kochakian, C. D., Gongora, J., and Parente, N.,** Metabolism of testosterone by homogenates of rabbit liver and kidney, *J. Biol. Chem.,* 196, 243, 1952.
29. **Kochakian, C. D. and Nall, D. M.,** The metabolism of epitestosterone by rabbit tissues in vitro, *J. Biol. Chem.,* 204, 91, 1953.
30. **Arimasa, N. and Kochakian, C. D.,** Epitestosterone and 5α-androstane-3α,17β-diol: the characteristic metabolites of androst-4-ene-3,17-dione produced by mouse kidney in vitro, *Endocrinology,* 92, 72, 1973.
31. **Short, R. V.,** Steroid concentrations in the follicular fluid of mares at various stages of the reproductive cycle, *J. Endocrinol.,* 22, 153, 1961.
32. **Dollefeld, E. and Breuer, H.,** Nachweis von Oestradiol: NAD-17α- and 17β-Oxidoreductasen in der Cytoplasmafraktion der Pferdeplacenta, *Biochim. Biophys. Acta,* 124, 187, 1966.
33. **Blaquier, J., Dorfman, R. I., and Forchielli, E.,** Formation of epitestosterone by human blood and adrenal tissue, *Acta Endocrinol.,* 54, 208, 1967.

34. **Linder, H. R.,** Androgens and related compounds in the spermatic vein blood of domestic animals. Species-linked differences in the metabolism of androstenedione in blood, *J. Endocrinol.*, 23, 161, 1961.
35. **Michel, F., Nicolas, J. C., and Crastes de Paulet, A.,** 17β-Hydroxysteroid dehydrogenase of the sheep ovary: purification, properties and substrate binding site, *Biochimie*, 57, 1131, 1975.
36. **Hussein, K. A. and Kochakian, C. D.,** DPN- and TPN-17β-hydroxy-C_{19}-steroid dehydrogenases; intracellular location in dog prostate, *Acta Endocrinol.*, 59, 459, 1968.
37. **Pollow, K., Lubbert, H., Boquoi, E., Kreutzer, G., Jeske, R., and Pollow, B.,** Studies on 17β-hydroxysteroid dehydrogenase in human endometrium and endometrial carcinoma. Subcellular distribution and variations of specific enzyme activity, *Acta Endocrinol.*, 79, 134, 1975.
38. **Pollow, K., Lubbert, H., and Pollow, B.,** On the mitochondrial 17β-hydroxysteroid dehydrogenase from human endometrium and endometrial carcinoma: characterization and intramitochondrial distribution, *J. Steroid Biochem.*, 7, 45, 1976.
39. **Pollow, K., Lubbert, H., Jeske, R., and Pollow, B.,** Studies of the 17β-hydroxysteroid dehydrogenase in human endometrium and endometrial carcinoma. Characterization of the soluble enzyme from secretory endometrium, *Acta Endocrinol.*, 79, 146, 1975.
40. **Pollow, K., Lubbert, H., and Pollow, B.,** Studies on 17β-hydroxysteroid dehydrogenase in human endometrium and endometrial carcinoma. Partial purification and characterization of the microsomal enzyme, *Acta Endocrinol.*, 80, 355, 1975.
41. **Endahl, G. L., Kochakian, C. D., and Hamm, D.,** Separation of a triphosphopyridine nucleotide-specific from a diphosphopyridine nucleotide-specific 17β-hydroxy-(testosterone) dehydrogenase of guinea pig liver, *J. Biol. Chem.*, 235, 2792, 1960.
42. **Liu, D. K. and Kochakian, C. D.,** Partial purification and some properties of guinea pig kidney 17β-hydroxy-C_{19} steroid dehydrogenase, *Steroids*, 19, 701, 1972.
43. **Littmann, K.-P., Gerdes, H., and Winter, G.,** Kinetik und Charakterisierung der oestradiolsensitiven 17β-Hydroxysteroiddehydrogenasen in der menschlichen Leber, *Acta Endocrinol.*, 67, 473, 1971.
44. **Tamaoki, B. and Inano, H.,** Testicular 17β-hydroxysteroid dehydrogenase: its distribution and properties, *J. Steroid Biochem.*, 6, 361, 1975.
45. **Karavolas, H. J. and Engel, L. L.,** Human placental 17β-estradiol dehydrogenase. Substrate specificity of the diphosphopyridine nucleotide(triphosphopyridine nucleotide)-linked enzyme, *Endocrinology*, 88, 1165, 1971.
46. **Findlay, J. K. and Breuer, H.,** Influence of an 18-hydroxyl group on the interaction of oestrogens and hydroxysteroid oxidoreductases, *Biochem. J.*, 137, 273, 1974.
47. **von Lehmann, W. D. and Breuer, H.,** Charakterisierung und Kinetik einer mikrosomalen 17β-Hydroxysteroid-Oxydoreduktase der menschlichen Placenta, *Hoppe-Seylers Z. Physiol. Chem.*, 348, 1633, 1967.
48. **Pollow, K., Sokolowski, G., Grunz, H., and Pollow, B.,** Hydroxysteroid oxidoreductases and their role in the catalysis of the specific transfer of hydrogen between steroid hormones in human placenta; morphological studies and the distribution of the activities of 17β- and 20α-hydroxysteroid oxidoreductases in subcellular fractions, *Hoppe-Seylers Z. Physiol. Chem.*, 355, 501, 1974.
49. **Thomas, C. M. G. and Veerkamp, J. H.,** The subcellular distribution of 17β-hydroxysteroid dehydrogenase in the human term placenta, *Acta Endocrinol.*, 82, 150, 1976.
50. **Isurugi, K., Inano, H., and Tamaoki, B.,** Submicrosomal distribution of human placental enzymes related to steroidogenesis, *Steroidologia*, 2, 303, 1971.
51. **Sweat, M. L., Bryson, M. J., and Young, R. B.,** Metabolism of 17β-estradiol and estrone by human proliferative endometrium and myometrium, *Endocrinology*, 81, 167, 1967.
52. **Tseng, L. and Gurpide, E.,** Estradiol and 20α-dihydroprogesterone dehydrogenase activities in human endometrium during the menstrual cycle, *Endocrinology*, 94, 419, 1974.
53. **Scublinsky, A., Marin, C., and Gurpide, E.,** Localization of estradiol 17β-dehydrogenase in human endometrium, *J. Steroid Biochem.*, 7, 745, 1976.
54. **Wenzel, M., Mutzel, W., and Hieronimus, B.,** Oestrous-cycle-dependent oestradiol oxidation in rat uterus and the influence of androst-3-en-17-one, *Biochem. J.*, 120, 899, 1970.
55. **Davenport, G. R. and Mallette, L. E.,** Some biochemical properties of rabbit ovarian hydroxysteroid dehydrogenases, *Endocrinology*, 78, 672, 1966.
56. **Lerner, N., Sulimovici, S., and Lunenfeld, B.,** Studies on 17β-hydroxysteroid dehydrogenase in immature mouse ovarian tissue, *J. Steroid Biochem.*, 4, 193, 1973.
57. **Inano, H. and Tamaoki, B.,** Relationship between steroids and pyridine nucleotides in the oxido-reduction catalysed by the 17β-hydroxysteroid dehydrogenase purified from the porcine testicular microsomal fraction, *Eur. J. Biochem.*, 53, 319, 1975.
58. **Villee, C. A. and Spencer, J. M.,** Some properties of the pyridine nucleotide-specific 17β-hydroxy steroid dehydrogenases of guinea pig liver, *J. Biol. Chem.*, 235, 3615, 1960.
59. **Endahl, G. L. and Kochakian, C. D.,** Partial purification and further characterization of the triphosphopyridine nucleotide specific C_{19}- 17β-hydroxysteroid dehydrogenase of guinea pig liver, *Biochim. Biophys. Acta*, 62, 245, 1962.

60. **Joshi, S. G., Duncan, E. L., and Engel, L. L.**, Soluble guinea pig liver TPN dependent 17β-hydroxysteroid (testosterone) dehydrogenase: partial purification and substrate specificity, *Steroids*, 1, 508, 1963.
61. **Kageura, E. and Toki, S.**, New aspect of guinea pig liver 17β-hydroxysteroid (testosterone) dehydrogenase, *Life Sci.*, 10, 469, 1971.
62. **Thaler-Dao, H., Descomps, B., Saintot, M., and Crastes de Paulet, A.**, The 17β-hydroxysteroid NAD(P) oxidoreductases from female rabbit liver cytosol: separation and characterization, *Biochimie*, 54, 83, 1972.
63. **Liu, D. K. and Kochakian, C. D.**, Heterogeneity of guinea pig kidney 17β-hydroxy-C_{19}-steroid dehydrogenase activity observed by disc gel electrophoresis, *Steroids*, 19, 721, 1972.
64. **Stevenson, D. and Kochakian, C. D.**, Purification of male guinea pig kidney 17β-hydroxy-C_{19}-steroid dehydrogenase, *Endocrinology*, 95, 766, 1974.
65. **Van der Molen, H. J. and Groen, D.**, Interconversion of progesterone and 20α-dihydroprogesterone and of androstenedione and testosterone in vitro by blood and erythrocytes, *Acta Endocrinol.*, 58, 419, 1968.
66. **Plotti, G., Menini, E., and Bompiani, A.**, Serum levels of oestradiol-17β dehydrogenase in normal and abnormal pregnancies, *J. Obstet. Gynaecol. Br. Commonw.*, 79, 603, 1972.
67. **Plotti, G., Menini, E., and Bompiani, A.**, Serum levels and biochemical characterization of oestradiol-17β-dehydrogenase in umbilical cord blood, *J. Endocrinol.*, 64, 103, 1975.
68. **Mulder, E., Lamers-Stahlohofen, G. J. M., and Van der Molen, H. J.**, Isolation and characterization of 17β-hydroxy steroid dehydrogenase from human erythrocytes, *Biochem. J.*, 127, 649, 1972.
69. **Lindner, H. R.**, The 17α-hydroxy-C_{19}-steroid dehydrogenase activity of ovine blood, *Steroids*, Suppl. 2, 133, 1965.
70. **Portius, H. J. and Repke, K.**, 17β-Estradiol dehydrogenase in rat erythrocytes, *Arch. Exp. Pathol. Pharmakol.*, 239, 144, 1960.
71. **Jacobsohn, G. M. and O'Rangers, J. J.**, Activation of 17β-hydroxysteroid dehydrogenase from human red blood cells, *Arch. Biochem. Biophys.*, 161, 384, 1974.
72. **Kato, J.**, In vitro uptake of tritiated estradiol by the rat anterior hypothalamus during the oestrous cycle, *Acta Endocrinol.*, 63, 577, 1970.
73. **Stumpf, W. E.**, Estrogen-neurons and estrogen-neuron systems in the periventricular brain, *Am. J. Anat.*, 129, 207, 1970.
74. **Harris, G. W. and Naftolin, F.**, The hypothalamus and control of ovulation, *Br. Med. Bull.*, 26, 3, 1970.
75. **Jaffe, R. B.**, Testosterone metabolism in target tissues: hypothalamic and pituitary tissues of the adult rat and human fetus, and the immature rat epiphysis, *Steroids*, 14, 483, 1969.
76. **Sholiton, L. J. and Werk, E. E.**, The less-polar metabolites produced by incubation of testosterone-4-^{14}C with rat and bovine brain, *Acta Endocrinol.*, 61, 641, 1969.
77. **Luttge, W. G. and Whalen, R. E.**, The accumulation, retention and interaction of oestradiol and oestrone in central neural and peripheral tissues of gonadectomized female rats, *J. Endocrinol.*, 52, 379, 1972.
78. **Knapstein, P., David, A., Wu, C. H., Archer, D. F., Flickinger, G. L., and Touchstone, J. C.**, Metabolism of free and sulfoconjugated DHEA in brain tissue in vivo and in vitro, *Steroids*, 11, 885, 1968.
79. **Perez-Palacios, G., Castaneda, E., Gomez-Perez, F., Perez, A. E., and Gual, C.**, In vitro metabolism of androgens in dog hypothalamus, pituitary and limbic system, *Biol. Reprod.*, 3, 205, 1970.
80. **Reddy, V. V., Naftolin, F., and Ryan, K. J.**, Steroid 17β-oxidoreductase activity in the rabbit central nervous system and adenohypophysis, *J. Endocrinol.*, 62, 401, 1974.
81. **Rommerts, F. F. G. and Van der Molen, H. J.**, Occurrence and localization of 5α-steroid reductase, 3α- and 17β-hydroxysteroid dehydrogenases in hypothalamus and other brain tissues of the male rat, *Biochim. Biophys. Acta*, 248, 489, 1971.
82. **Weinstein, G. D., Frost, P., and Hsia, S. L.**, In vitro interconversion of estrone and 17β-estradiol in human skin and vaginal mucosa, *J. Invest. Dermatol.*, 51, 4, 1968.
83. **Korenman, S. G., Wilson, H., and Lipsett, M. B.**, Isolation of 17α-hydroxyandrost-4-ene-3-one (epitestosterone) from human urine, *J. Biol. Chem.*, 239, 1004, 1964.
84. **Brooks, R. V. and Guiliani, G.**, Epitestosterone: isolation from human urine and experiments on possible precursors, *Steroids*, 4, 101, 1964.
85. **Sparagana, M.**, Quantitative gas chromatographic analysis of urinary testosterone and epitestosterone, *Steroids*, 5, 773, 1965.
86. **Futterweit, W., Freeman, R., Siegel, G. L., Griboff, S. I., Dorfman, R. I., and Soffer, L. J.**, Chemical applications of a gas chromatographic method for the combined determination of testosterone and epitestosterone glucuronide in urine, *J. Clin. Endocrinol. Metab.*, 25, 1451, 1965.
87. **Hasnain, S. and Williamson, D. G.**, Properties of the multiple forms of the soluble 17α-hydroxy steroid dehydrogenase of rabbit liver, *Biochem. J.*, 161, 279, 1977.

88. **Hagerman, D. D.,** Monomer-trimer transition of human placenta estradiol dehydrogenase, *Arch. Biochem. Biophys.*, 134, 196, 1969.
89. **Jarabak, J. and Sack, G. H.,** A soluble 17β-hydroxysteroid dehydrogenase from human placenta. The binding of pyridine nucleotides and steroids, *Biochemistry*, 8, 2203, 1969.
90. **Burns, D. J. W., Engel, L. L., and Bethune, J. L.,** The subunit structure of human placental 17β-estradiol dehydrogenase, *Biochem. Biophys. Res. Commun.*, 44, 786, 1971.
91. **Jarabak, J. and Street, M. A.,** Studies on the soluble 17β-hydroxysteroid dehydrogenase from human placenta. Evidence for a subunit structure, *Biochemistry*, 10, 3831, 1971.
92. **Pollow, K., Lubbert, H., and Pollow, B.,** Partial purification and evidence of heterogeneity of the cytoplasmic 17β-hydroxysteroid dehydrogenase from normal human endometrium and endometrial carcinoma, *J. Steroid Biochem.*, 7, 315, 1976.
93. **Kochakian, C. D., Stevenson, D., and Mayumi, T.,** Dependence of the guinea pig kidney 17β-hydroxy-C_{19}-steroid dehydrogenase isoenzyme profile on androgen, *Biochem. Biophys. Res. Commun.*, 54, 519, 1973.
94. **Branden, C. I., Jornvall, H., Eklund, H., and Furugren, B.,** Alcohol dehydrogenases, in *The Enzymes*, Vol. XI, Boyer, P. D., Ed., Academic Press, New York, 1975, 103.
95. **Warren, J. C. and Crist, R. D.,** Site specificity and mechanism of human placental 17β-hydroxysteroid dehydrogenase, *Arch. Biochem. Biophys.*, 118, 577, 1967.
96. **Betz, G.,** Reaction mechanism of 17β-estradiol dehydrogenase determined by equilibrium rate exchange, *J. Biol. Chem.*, 246, 2063, 1971.
97. **Jarabak, J. and Talalay, P.,** Stereospecificity of hydrogen transfer by pyridine nucleotide-linked hydroxysteroid dehydrogenases, *J. Biol. Chem.*, 325, 2147, 1960.
98. **Achtar, M., Wilton, D. C., Watkinson, I. A., and Rachimtula, A. D.,** *Proc. R. Soc. Lond. Ser. B.*, 180, 167, 1972.
99. **George, J. M., Orr, J. C., Renwick, A. G. C., Carter, P., and Engel, L. L.,** The stereochemistry of hydrogen transfer to NADP by enzymes acting upon stereoisomeric substrates, *Bioorg. Chem.*, 2, 140, 1973.
100. **Groman, E. V., Schultz, R. M., Engel, L. L., and Orr, J. C.,** Horse liver alcohol dehydrogenase and *Pseudomonas testosteroni* 3(17)β-hydroxysteroid dehydrogenase transfer epimeric hydrogens from NADH to 17β-hydroxy-5α-androstan-3-one, *Eur. J. Biochem.*, 63, 427, 1976.
101. **Biellmann, J. C., Branlant, G., Nicolas, J. C., Pons, M., Descomps, B., and Crastes de Paulet, A.,** Alkylation of estradiol 17β-dehydrogenase from human placenta with 3-chloroacetylpyridine-adenine dinucleotide, *Eur. J. Biochem.*, 63, 477, 1976.
102. **Nicolas, J. C. and Harris, J. I.,** Human placental 17β-oestradiol dehydrogenase. Sequence of a tryptic peptide containing an essential cysteine, *FEBS Lett.*, 29, 173, 1973.
103. **Pons, M., Nicolas, J. C., Boussioux, A. M., Descomps, B., and Crastes de Paulet, A.,** Affinity labelling of an histidine of the active site of the human placental 17β-oestradiol dehydrogenase, *FEBS Lett.*, 36, 23, 1973.
104. **Pons, M., Nicolas, J. C., Boussioux, A. M., Descomps, B., and Crastes de Paulet, A.,** Affinity labelling of the estradiol-17β dehydrogenase from human placenta with substrate analogs, *Eur. J. Biochem.*, 68, 385, 1976.
105. **Groman, E. V., Schultz, R. M., and Engel, L. L.,** Catalytic competence, a new criterion for affinity labelling, *J. Biol. Chem.*, 250, 5450, 1975.
106. **Bongiovanni, A. M. and Marino, J.,** Steroid dehydrogenases in rabbit liver of various developmental stages, *J. Steroid Biochem.*, 3, 859, 1972.
107. **Ghraf, R., Vetter, U., Zandveld, J. M., and Schriefers, H.,** Organ-specific ontogeneses of steroid hormone metabolizing enzyme activities in the rat, *Acta Endocrinol.*, 79, 192, 1975.
108. **Baldi, A. and Charreau, E. H.,** 17β-Hydroxysteroid dehydrogenase activity in rat submaxillary glands. Its relation to sex and age, *Endocrinology*, 90, 1643, 1972.
109. **Schriefers, H., Hoff, H. G., and Ghraf, R.,** Androgen dependency of the ontogenesis of the activity patterns of enzymes involved in the metabolism of steroid hormones in the rat liver, *Hoppe-Seylers Z. Physiol Chem.*, 354, 501, 1973.
110. **Denef, C.,** Differentiation of steroid metabolism in the rat and mechanisms of neonatal androgen action, *Enzyme*, 15, 254, 1973.
111. **Einarsson, K., Gustafsson, J. A., and Stenberg, A.,** Neonatal imprinting of liver microsomal hydroxylation and reduction of steroids, *J. Biol. Chem.*, 248, 4987, 1973.
112. **Gustaffson, J. A. and Stenberg, A.,** Irreversible androgenic programming at birth of microsomal and soluble rat liver enzymes active on 4-androstene -3, 17-dione and 5α-androstane-3α, 17β-diol, *J. Biol. Chem.*, 249, 711, 1974.
113. **Kraulis, I. and Clayton, R. B.,** Sexual differentiation of testosterone metabolism exemplified by the accumulation of 3β, 17α-dihydroxy-5α-androstane 3-sulfate as a metabolite of testosterone in the castrated rat, *J. Biol. Chem.*, 243, 3546, 1968.

114. **de Nicola, A. F., Dorfman, R. I., and Forchielli, E.**, Urinary excretion of epitestosterone and testosterone in normal individuals and hirsute and virilized females, *Steroids*, 7, 351, 1966.
115. **Lax, E. R., Hoff, H. G., Ghraf, R., Schroder, E., and Schriefers, H.**, The role of the hypophysis in the regulation of sex differences in the activities of enzymes involved in hepatic steroid hormone metabolism, *Hoppe-Seylers Z. Physiol. Chem.*, 355, 1325, 1974.
116. **Ghraf, R., Lax, E. R., and Schriefers, H.**, The hypophysis in the regulation of androgen and oestrogen dependent enzyme activities of steroid hormone metabolism in rat liver cytosol, *Hoppe-Seylers Z. Physiol. Chem.*, 356, 127, 1975.
117. **Ghraf, R., Lax, E. R., Hoff, H. G., and Schriefers, H.**, The role of the gonads and the hypophysis in the regulation of hydroxysteroid dehydrogenase activities in rat kidney, *Hoppe-Seylers Z. Physiol. Chem.*, 356, 135, 1975.
118. **Musto, N., Hafiez, A. A., and Bartke, A.**, Prolactin increases 17β-hydroxysteroid dehydrogenase activity in the testis, *Endocrinology*, 91, 1106, 1972.
119. **Jensen, E. V. and Jacobson, H. I.**, Basic guides to the mechanism of estrogen action, in *Recent Progress in Hormone Research*, Vol. 18, Pincus, G., Ed., Academic Press, New York, 1962, 387.
120. **MacCartney, J. C. and Thomas, G. H.**, NADP-linked 17β- and 20α-steroid reductase activity in the rabbit uterus, *J. Endocrinol.*, 43, 247, 1969.
121. **Pack, B. A. and Brooks, S. C.**, Metabolism of estrogen and their sulfates in rat uterine minces, *Endocrinology*, 87, 924, 1970.
122. **Jutting, G., Thun, K. J., and Kuss, E.**, Purification and characterization of estrogen induced 17β-hydroxy-steroid NAD(P)-oxidoreductase from the myometrium in rabbits, *Eur. J. Biochem.*, 2, 146, 1967.
123. **Oshima, H., Fan, D. F., and Troen, P.**, Studies of the human testis. Properties of Δ5-3β and 17β-hydroxysteroid dehydrogenases in the biosynthesis of testosterone from dehydroepiandrosterone, *J. Clin. Endocrinol. Metab.*, 40, 573, 1975.
124. **Sulimovici, S. and Lunenfeld, B.**, The effect of adenosine 3′,5′-cyclic-monophosphoric acid on the 17β-hydroxysteroid dehydrogenase of rat testis, *J. Steroid Biochem.*, 3, 781, 1972.
125. **Murad, F., Strauch, B. S., and Vaughan, M.**, The effect of gonadotropins on testicular adenyl cyclase, *Biochem. Biophys. Acta*, 177, 591, 1969.
126. **Kuehl, F. A. Jr., Patanelli, D. J., Tarnoff, J., and Humes, J. L.**, Testicular adenyl cyclase: stimulation by the pituitary gonadotrophins, *Biol. Reprod.*, 2, 154, 1970.
127. **Shaw, M. A. and Jeffery, J.**, Inhibition of human placental oestradiol-17β dehydrogenase by adenosine triphosphate: the importance of the coenzyme, *Biochem. Soc. Trans.*, 4, 1097, 1976.
128. **Saldarini, R. J., Hilliard, J., Abraham, G. E., and Sawyer, C. H.**, Relative potencies of 17α- and 17β-estradiol in the rabbit, *Biol. Reprod.*, 3, 105, 1970.
129. **Baird, D., Horton, R., Longcope, C., and Tait, J. F.**, Steroid prehormones, *Perspect. Biol. Med.*, 11, 384, 1968.
130. **Russell, S. L. and Thomas, G. H.**, The interconversion of oestrone and oestradiol-17β and the concentration of oestradiol-17β-receptors in cultured rabbit uterus, *J. Endocrinol.*, 56, 203, 1973.
131. **Crocker, J. M. and Thomas, G. H.**, Uptake and interconversion of tritiated oestrone and oestradiol-17β in the intact rabbit in vivo, *J. Endocrinol.*, 56, 213, 1973.
132. **Tseng, L., Tseng, J. K., Escarcena, L., and Gurpide, E.**, Comparison of results from in situ and in vitro perfusions of rabbit uterus with labeled estrone and estradiol, *Endocrinology*, 97, 1481, 1975.
133. **Raspi, G., Ed.**, *Advances in the Biosciences 7*, Schering Workshop on Steroid Hormone Receptors, Pergamon Press, Oxford, 1971.
134. **Heald, P. J. and McLachlan, P. M.**, The synthesis of phosvitin in vitro by slices of liver from the laying hen, *Biochem. J.*, 94, 32, 1965.
135. **Greengard, O., Sentenac, A., and Acs, G.**, Induced formation of phosphoprotein in tissues of cockerels in vivo and in vitro, *J. Biol. Chem.*, 240, 1687, 1965.
136. **Arias, F. and Warren, J. C.**, An estrophilic macromolecule in chicken liver cytosol, *Biochim. Biophys. Acta*, 230, 550, 1971.
137. **Gschwendt, M.**, A cytoplasmic oestrogen-binding component in chicken liver, *Hoppe-Seylers Z. Physiol. Chem.*, 356, 157, 1975.
138. **Van Beurden-Lamers, W. M. O., Brinkmann, A. O., Mulder, E., and Van der Molen, H. J.**, High-affinity binding of oestradiol-17β by cytosols from testis interstitial tissue, pituitary, adrenal, liver and accessory sex glands of the male rat, *Biochem. J.*, 140, 495, 1974.
139. **Layne, D. S., Sheth, N. A., and Kirdani, R. Y.**, The isolation from rabbit urine of a conjugate of 17α-estradiol with *N*-acetylglucosamine, *J. Biol. Chem.*, 239, 3221, 1964.
140. **Layne, D. S.**, Identification in rabbit urine of the 3-glucuronoside-17-*N*-acetylglucosaminide of 17α-estradiol, *Endocrinology*, 76, 600, 1965.
141. **Milgrom, E. and Baulieu, E. E.**, C_{19}-steroid conjugates as substrates and inhibitors of 17β-hydroxysteroid oxidoreductases, *Steroids*, 15, 563, 1970.

142. **Hobkirk, R., Nilsen, M., and Jennings, B.,** 17-Oxidoreduction of 17β-estradiol, estrone and their 3-sulfates by kidney slices from guinea pig and human, *Can. J. Biochem.*, 53, 1333, 1975.
143. **Robel, P., Emiliozzi, R., and Baulieu, E. E.,** Studies on testosterone metabolism. The selective "5β-metabolism" of testosterone glucuronide, *J. Biol. Chem.*, 241, 20, 1966.
144. **Villee, C. A. and Hagerman, D. D.,** Effects of estradiol on the metabolism of human placenta in vitro, *J. Biol. Chem.*, 205, 873, 1953.
145. **Talalay, D. and Williams-Ashman, H. G.,** Activation of hydrogen transfer between pyridine nucleotides by steroid hormones, *Proc. Natl. Acad. Sci. U.S.A.*, 44, 15, 1958.
146. **Talalay, P., Hurlock, B., and Williams-Ashman, H. S.,** On a coenzyme function of estradiol-17β, *Proc. Natl. Acad. Sci. U.S.A.*, 44, 862, 1958.
147. **Jarabak, J., Adams, J. A., Williams-Ashman, H. G., and Talalay, P.,** Purification of a 17β-hydroxysteroid dehydrogenase of human placenta and studies on its transhydrogenase function, *J. Biol. Chem.*, 237, 345, 1962.
148. **Hagerman, D. D. and Villee, C. A.,** Separation of human placental estrogen-sensitive transhydrogenase from estradiol-17β dehydrogenase, *J. Biol. Chem.*, 234, 2031, 1959.
149. **Hagerman, D. D. and Villee, C. A.,** Kinetics of nicotinamide nucleotide transhydrogenation by human placenta, *Biochem. J.*, 95, 150, 1965.
150. **Karavolas, H. J., Orr, J. C., and Engel, L. L.,** Human placental 17β-estradiol dehydrogenase. Differentiation of 17β-estradiol-activated transhydrogenase from the transhydrogenase function of 17β-estradiol dehydrogenase, *J. Biol. Chem.*, 244, 4413, 1969.
151. **Abe, T., Hagerman, D. D., and Villee, C. A.,** Estrogen-dependent pyridine nucleotide transhydrogenase of human myometrium, *J. Biol. Chem.*, 239, 414, 1964.
152. **Thompson, E. A. and Siiteri, P. K.,** Utilization of oxygen and reduced nicotinamide adenine dinucleotide phosphate by human placental microsomes during aromatization of androstenedione, *J. Biol. Chem.*, 249, 5364, 1974.
153. **Wenzel, M. and Mutzel, W.,** The influence of hormonally inactive steroids on the oxidation of oestradiol: studies on the transfer of hydrogen from oestradiol-17β to 5α-androsten-17-one in rat tissue, *Hoppe-Seylers Z. Physiol. Chem.*, 351, 1221, 1970.
154. **Pollow, K., Sokolowski, G., Schmalbeck, J., and Pollow, B.,** Hydroxysteroid oxidoreductases and their role in the catalysis of the specific transfer of hydrogen between steroid hormones in human placenta. Kinetic studies on the hydrogen transfer between C-17 of estradiol and C-20 of progesterone, *Hoppe-Seylers Z. Physiol. Chem.*, 355, 515, 1975.
155. **Beaulieu, E. E.,** Some aspects of the mechanism of action of steroid hormones, *Mol. Cell. Biochem.*, 7, 157, 1975.
156. **Edelman, I. S.,** Mechanism of action of steroid hormones, *J. Steroid Biochem.*, 6, 147, 1975.
157. **Pietras, R. J. and Szego, C. M.,** Specific binding sites for oestrogen at the outer surfaces of isolated endometrial cells, *Nature*, 265, 69, 1977.
158. **Watanabe, M. and Po, L.,** Testosterone uptake by membrane vesicles of *Pseudomonas testosteroni*, *Biochim. Biophys. Acta*, 345, 419, 1974.
159. **Marcus, P. I. and Talalay, P.,** Induction and purification of α- and β-hydroxysteroid dehydrogenases, *J. Biol. Chem.*, 218, 661, 1956.
160. **Watanabe, M. and Po, L.,** Membrane bound 3β and 17β-hydroxysteroid dehydrogenase and its role in steroid transport in membrane vesicles of *Pseudomonas testosteroni*, *J. Steroid Biochem.*, 7, 171, 1976.
161. **Lefebvre, Y., Po, L., and Watanabe, M.,** Effect of sulfhydryl and disulfide agents on 3β and 17β-hydroxysteroid dehydrogenases and on steroid uptake of Pseudomonas testosteroni, *J. Steroid Biochem.*, 7, 171, 1976.
162. **Lefebvre, Y., Po, L., and Watanabe, M.,** The involvement of the electron transport chain in uptake of testosterone by membrane vesicles of *Pseudomonas testosteroni*, *J. Steroid Biochem.*, 7, 867, 1976.

Chapter 4

THE ROLE OF STEROID SULFATASE AND SULFOTRANSFERASE ENZYMES IN THE METABOLISM OF C_{21} AND C_{19} STEROIDS

A. H. Payne and S. S. Singer

TABLE OF CONTENTS

I. INTRODUCTION

Considerable interest in the biological role of steroid sulfates has developed during the past decade. It is now well accepted that steroid sulfates are not merely end products of metabolism whose only fate is removal via excretion. Steroid sulfates are found in high concentrations in human plasma[1] (Table 1); they can serve as intermediates in steroid metabolism[2-4] and as precursor pools of biologically active androgens[5] and estrogens.[6] Glucocorticoid sulfates have been implicated in certain types of hypertension and in the induction of hepatic tyrosine transaminase.

The concentration and tissue distribution of steroid sulfates are dependent on two classes of enzymes: (1) steroid sulfotransferases that catalyze the formation of these compounds and (2) steroid sulfatases that catalyze their hydrolysis to free steroids. This chapter is devoted to a discussion of these enzymes as they affect the metabolism of C_{19} and C_{21} steroids. It has been divided into two major categories. The first includes the metabolism of the sulfate esters of 3β-hydroxy C_{21} and C_{19} steroids; the second concentrates on the formation and potential metabolic role of glucocorticoid sulfates. Other chapters in this book will discuss different aspects of steroid sulfate conjugation.

II. C_{21} AND C_{19} 3β-HYDROXYSTEROID SULFATES

A. Steroid Sulfotransferases

The biosynthesis of steroid sulfates is brought about in two distinct reactions. The first is the formation of activated sulfate, 3′-phosphoadenosine-5′phosphosulfate (PAPS), in cytoplasm. This involves two separate steps, carried out by the enzymes sulfate adenylyl transferase (E.C. 2.7.7.4.) and adenylyl sulfate kinase (E.C. 2.7.1.25), which catalyze the formation of adenosine-5′-phosphosulfate (APS) and PAPS as shown below.

$$\text{ATP} + \text{SO}_4^{\ 2-} \underset{}{\overset{\text{Mg}^{2+}}{\rightleftharpoons}} \text{APS} + \text{PP}_i$$

$$\text{APS} + \text{ATP} \xrightarrow{\text{Mg}^{2+}} \text{PAPS} + \text{ADP}$$

The second reaction is the transfer of the activated sulfate by steroid sulfotransferases to a steroid. (Steroid sulfotransferases are sometimes incorrectly referred to as sulfokinases.)

$$\text{Steroid} - \text{OH} + \text{PAPS} \xrightarrow{\text{sulfotransferase}} \text{Steroid} - \text{O} - \text{SO}_3^{\ -} + \text{PAP}$$

The steroid sulfotransferases are soluble enzymes. To date, two such enzymes have been recognized by the Commission on Biochemical Nomenclature.[7] The first, 3′-phosphoadenylylsulfate: oestrone 3′-sulfotransferase (E.C. 2.8.2.4.), is specific for the transfer of activated sulfate to the 3-hydroxy group of estrogens. The other, 3′-phosphoadenylylsulphate:3β-hydroxysteroid sulfotransferase (E.C. 2.8.2.2.), is specific for the transfer of the activated sulfate to the 3β-hydroxy group of C_{19} or C_{21} steroids. However, evidence for more than two steroid sulfotransferases is presented later in this chapter, in the section on glucocorticoid sulfates.

1. Liver

Most studies of steroid sulfotransferases have been conducted using liver as the enzyme source. Nose and Lipmann[8] were the first to separate two steroid sulfotransferases from rabbit liver. One catalyzed the formation of dehydroepiandrosterone sulfate

TABLE 1

Concentration of Steroid Sulfates in Plasma from Male and Female Blood Donors (Values are the Mean ± Standard Deviation, Expressed as μg Free Steroid in 100 mℓ Plasma)[a]

Sex[b]	Dehydroe-piandroster-one	Androstene-diol	Pregn-5-ene-3β-20α-diol	Androsterone	Epiandros-terone
M (62)	99.7 ± 65	28.7 ± 26.5	20.3 ± 11.3	30.6 ± 24.3	11.4 ± 7.4
F (54)	60.4 ± 40.3	11.3 ± 8.2	13.3 ± 10.2	13.6 ± 13.1	6.0 ± 4.1

[a] Data taken from Vihko.[1]
[b] M represents the mean of 62 males aged 20 to 70 years; F represents the mean of 54 females aged 20 to 70 years.

(3β-hydroxysteroid sulfotransferase) and the other catalyzed the formation of estrone sulfate (estrone-3-sulfotransferase). Both were distinct from the phenol sulfotransferase (E.C. 2.8.2.1.). The 3β-hydroxysteroid sulfotransferase also sulfated pregnenolone or epiandrosterone, but exhibited only minor activity towards other steroids. Banerjee and Roy[9] later resolved 3β-hydroxysteroid sulfotransferase and estrone-3-sulfotransferase from guinea pig liver. Both of these enzymes also sulfated *p*-nitrophenol. The guinea pig liver 3β-hydroxysteroid sulfotransferase sulfated dehydroepiandrosterone much more extensively than other steroids tested. Banerjee and Roy[9] also presented evidence that deoxycorticosterone was sulfated by an enzyme distinct from the 3β-hydroxysteroid sulfotransferase.

A recent study by Ryan and Carroll[10] described the 60-fold purification of a 3β-hydroxysteroid sulfotransferase from livers of adult female rats. Isolation from a 105,000 × g supernatant was accomplished using 25% ethanol precipitation followed by DEAE-cellulose and phosphocellulose chromatography. The partially purified enzyme exhibited its greatest activity with epiandrosterone, followed by dehydroepiandrosterone, 5α-pregnane-3β,20α-diol,5α-androstane-3β,17β-diol. It was less active with androst-5-ene-3β,17β-diol, pregnenolone, and androsterone and had almost no enzymatic activity with other steroids tested.

2. *Adrenal*

In addition to their localization in liver, C_{19} and C_{21} steroid sulfotransferases are found in significant amounts in adrenals, in testes, and in ovaries of adult mammals. Unlike the liver enzymes that appear to be found in all species, there is considerable species variation in the presence of steroid sulfotransferases in other tissues. Adams[11] demonstrated the presence of sulfotransferase activity in a 100,000 × g supernatant fraction obtained from fresh human female adrenal tissue. This soluble fraction exhibited its greatest activity with dehydroepiandrosterone, and somewhat less activity with 3α- and 3β-hydroxy, 17-keto and 5α- and β- reduced C_{19} steroids. Pregnenolone, testosterone, deoxycorticosterone, and estrone were also sulfated, but to a lesser degree. In a subsequent study, Adams and Edwards[12] reported kinetic studies of the partially purified human adrenal 3β-hydroxysteroid sulfotransferase. It showed anomolous kinetic properties which were attributed to its association into dimers, trimers, or higher associated states. The degree of association was related to enzymic activity and was dependent on the concentration of enzyme, substrate, and cofactors. They suggested that the enzyme possessed many of the characteristics of a regulatory enzyme and thus might be involved in determining the extent of steroid sulfate formation and secretion by the human adrenal gland. Holcenberg and Rosen[13] demonstrated the sulfation of dehydroepiandrosterone, but not of testosterone with 100,000 × g supernatant from bovine adrenals. Fry and Koritz[14] reported the absence of pregnenolone sulfotransferase activity with a soluble fraction from rat adrenals. (As described later in this chapter, the authors' laboratory could not demonstrate dehydroepiandrosterone sulfotransferase activity in a soluble preparation from rat adrenals.) Anderson et al.[15] were not able to detect pregnenolone sulfate or dehydroepiandrosterone sulfate after incubations of [^{3}H] pregnenolone with chopped adrenal glands from pregnant ewes.

The observation that dehydroepiandrosterone sulfate is the major secretory product of human adrenal[16,17] has been disputed in two recent studies.[18,19] Data reported by Nieschlag et al.[18] and by Doouss et al.[19] suggest that although the human adrenal secretes dehydroepiandrosterone sulfate, dehydroepiandrosterone secretion is quantitatively greater than that of its sulfate conjugate. The biological implications of dehydroepiandrosterone sulfate secretion by the human adrenal gland remain to be determined.

3. Gonads

Although investigations on steroid sulfotransferases in gonads are very limited, there are several studies on secretion and tissue concentration of gonadal steroid sulfates that indicate, especially in the testis, the presence of steroid sulfotransferases in these tissues. These studies suggest that steroid sulfates may play a major role in steroid metabolism in the testis of some species.

a. Ovaries

The formation of the 3-monosulfates of dehydroepiandrosterone and Δ^5-androstenediol by minces and homogenates of normal human ovaries was demonstrated by Wallace and Silberman.[20] Holcenberg and Rosen[13] were unable to detect any formation of dehydroepiandrosterone sulfate upon incubation of homogenates of bovine ovaries with dehydroepiandrosterone and appropriate cofactors. Sandberg et al.[21] observed some dehydroepiandrosterone sulfation in incubations of human ovarian arrhenoblastomatous tissue slices. Kallialla et al.[22] were unable to demonstrate ovarian secretion of steroid sulfates in a study that compared the concentration of various C_{21} and C_{19} steroid sulfates in peripheral plasma to their concentration in ovarian vein blood of women. It thus appears that the formation of C_{21} or C_{19} steroid sulfates in human and bovine ovaries is minimal or nonexistent.

b. Testes

Dixon et al.[23] demonstrated the formation of dehydroepiandrosterone sulfate and testosterone sulfate after incubation of either dehydroepiandrosterone or testosterone with minced testicular tissue from a 19-year-old male patient. Pierrepoint et al.[24] observed the sulfation of pregnenolone, 17-hydroxypregnenolone, and dehydroepiandrosterone by minces and homogenates of a feminizing testicular Leydig cell tumor. Pérez-Palacios et al.[25] demonstrated the synthesis of pregnenolone sulfate, 17-hydroxypregnenolone sulfate, and dehydroepiandrosterone sulfate from [^{14}C] acetate in testicular tissue slices from two patients with the complete form of testicular feminization. This indicates that human testicular tissue can synthesize C_{21} and C_{19} steroid sulfates *de novo*.

Vihko and co-workers[26] have measured free steroid and steroid monosulfate concentrations in human testes obtained at the time of orchiectomy for prostatic carcinoma or at the time of autopsy. Samples obtained from cadaver testes and at the time of orchiectomy had similar steroid compositions. With the exception of testosterone, pregnenolone sulfate was quantitatively the major steroid found in human adult testes (Table 2). In addition, two other steroids that could serve as precursors of testosterone were present mostly as sulfate conjugates. On the other hand, testosterone that is se-

TABLE 2

Concentration of Steroids and Steroid Monosulfates in Human Testes (Values are Mean Concentrations Expressed in μg Free Steroid per 100 g Tissue)[a]

Steroid	Free steroid	Steroid monosulfate
Pregnenolone	25	51
Dehydroepiandrosterone	3	41
5-Androstene-3β,17β-diol	3	19
Testosterone	55	6.8

[a] Testes were obtained at the time of orchiectomy for prostatic carcinoma. Data taken from Ruokonen et al.[26]

creted by the testis as the free steroid and that is essential for completion of spermatogenesis[27] was found predominantly in the unconjugated form. The same group of investigators carried out additional studies on the secretion of steroid sulfates by human testes and on the possible influence of gonadotropins on the formation of steroid sulfates in human testes. Laatikainen et al.[28] measured steroid sulfate concentration in spermatic vein plasma and peripheral plasma of normal subjects. They demonstrated that pregnenolone sulfate, small amounts of 5-androstene-3β,17β-diol, and testosterone sulfate are secreted by the human testis (Table 3). In a subsequent study,[29] they found that considerable amounts of pregnenole sulfate, dehydroepiandrosterone sulfate, and 5-androstene-3β,17β-diol-3-sulfate were secreted by testes of three patients after stimulation with hCG. Under these conditions, testosterone was secreted almost exclusively as the free steroid. It cannot be ascertained whether the increased secretion of the Δ^5-C_{21} and C_{19} steroid sulfates was a direct effect of hCG on sulfation of these steroids or whether increased formation of free steroids occurred, followed by sulfation.

The secretion of dehydroepiandrosterone sulfate by the boar testis has been reported by Baulieu et al.[30] These investigators measured the concentrations of dehydroepiandrosterone, dehydroepiandrosterone sulfate, testosterone, and androstenedione in boar testes and spermatic vein plasma after treatment with hCG for 3 days. Dehydroepiandrosterone sulfate and unconjugated testosterone were the most abundant steroids in both testicular tissue and spermatic vein plasma. In addition to dehydroepiandrosterone sulfate, Ruokonen and Vihko[31] reported the presence of other C_{19} and C_{21} steroid sulfates in boar testes, including the 3β-monosulfates of 5-androstene-3β,17β-diol of 5α-androstane-3β,17β-diol and of 5-pregnen-3β,20α-diol.

Rat testes exhibit a very different pattern of endogenous steroids from those of human[26] or boar[31] testes. Ruokonen et al.,[32] using extraction and identification methods reported in their studies of human and boar testes,[29,32] were unable to detect any steroid sulfates in rat testicular tissue. In addition, testosterone and pregnenolone concentrations (μg/100 g tissue wet weight) were considerably lower than those in human testes: 10.6 and 0.8 μg, respectively, compared to 55 and 25 μg in human testicular tissue.[27]

Studies presently in progress in the authors' laboratory[33] on steroid sulfotransferase in testicular compartments of rats demonstrate that steroid sulfotransferase activity is found exclusively in seminiferous tubules. Rat testis interstitial tissue was separated from seminiferous tubules by a method involving treatment with collagenase followed by a wet dissection procedure. The 105,000 × g soluble fraction from each testicular compartment was subjected to $(NH_4)_2SO_4$ fractionation. The 50% $(NH_4)_2SO_4$ precip-

TABLE 3

Concentrations of Steroid Sulfates in Spermatic Vein and Peripheral Vein Plasma (Expressed as μg Free Steroid/100 m*l* Plasma)[a]

			Steroid Sulfates			
Patient	Age	Sample[b]	Pregnenolone	Dehydroepiandrosterone	5-Androstene-3β,17β-diol	Testosterone
1	30	P	7.2	103	9.1	—
		S	23.0	83	10.0	2.8
2	30	P	8.7	62	5.6	—
		S	14.0	87	9.3	1.7
3	37	P	11.0	124	7.0	—
		S	20.0	102	7.4	1.2
4	39	P	6.1	53	4.3	—
		S	32.0	52	5.2	3.2
5	65	P	3.2	30	4.1	—
		S	29.0	41	7.3	3.9
6	67	P	5.0	53	2.3	—
		S	35.4	78	12.0	1.8
7	67	P	4.1	28	2.2	—
		S	5.9	23	4.0	0.9
8	74	P	2.6	29	5.2	—
		S	17.0	40	6.8	1.6
9	83	P	2.8	21	2.9	—
		S	5.8	20	3.6	<1

[a] Data taken from Laatikainen, et al.[28].
[b] P = peripheral vein; S = spermatic vein.

itate was dissolved in water, dialyzed, and centrifuged to remove nonsoluble material and the supernatant used as the source of sulfotransferase. Appropriate amounts of this fraction were incubated with [^{3}H] dehydroepiandrosterone and a saturating concentration of 3′-phosphoadenosine-5′-phosphosulfate (0.1 m*M*). No sulfotransferase activity was detected in isolated interstitial tissue. Approximately 12 pmol of dehydroepiandrosterone sulfate was synthesized per 100 min/mg protein with enzyme preparations from isolated seminiferous tubules. Enzyme preparations from rat liver formed about ten times as much dehydroepiandrosterone sulfate per milligram protein as preparations from testes. No dehydroepiandrosterone sulfate was detected in incubations with adrenal tissue preparations. Additional studies with enzyme preparations from isolated seminiferous tubules of rat testes showed that the K_m for dehydroepiandrosterone was 3 μ*M* and the pH optimum was 10. When dehydroepiandrosterone was incubated in the presence of equimolar concentrations of various other steroids, pregnenolone was the most potent inhibitor. Pregnenolone acted as a competitive inhibitor, exhibiting a K_i of approximately 1/10 the K_m for dehydroepiandrosterone. These results are in contrast to the reports for the 3β-hydroxysteroid sulfotransferase from rat liver or human adrenal,[11] in which it was reported that dehydroepiandrosterone was the preferred substrate.

4. Fetal Tissues

Steroid sulfates are present in relatively high concentration in the human fetal circulation. Conrad et al.[34] measured pregnenolone sulfate in maternal and fetal plasma and found that concentrations were higher in the latter. The major site of neutral steroid sulfate formation in the human fetus appears to be the adrenal gland. Wengle[35] investigated steroid sulfotransferase activity in the 105,000 × g supernatant of homogenates from a number of fetal tissues with a variety of C_{19}, C_{21}, and C_{18} steroids and ^{35}S-labeled PAPS. Fetal adrenal extracts exhibited approximately 30 times the specific activity of liver extracts expressed per gram of wet weight. Dehydroepiandrosterone, etiocholan-3α-ol-17-one, and pregnan-3β-ol-20-one were sulfated to the greatest extent in both tissues. Estrone sulfation was only one sixth that of dehydroepiandrosterone in adrenal extracts. Liver extracts exhibited approximately the same capacity for sulfation of dehydroepiandrosterone as for estrone. Kidney and jejunum extracts were the only other tissues which exhibited steroid sulfotransferase activity toward dehydroepiandrosterone. Huhtaniemi et al.[36] studied the concentration of a number of C_{21} and C_{19} steroid sulfates in human fetal adrenals and livers. They found that fetal adrenals contained much larger amounts of steroid sulfates than did fetal livers. They also observed that fetal adrenals contained five times as much pregnenolone sulfate as dehydroepiandrosterone sulfate. Jaffe et al.[37] reported the *de novo* synthesis of pregnenolone sulfate, 17-hydroxypregnenolone sulfate, and dehydroepiandrosterone sulfate from [^{14}C] acetate in slices of human fetal adrenals. Huhtaniemi[38] determined the concentration of various free steroids and steroid sulfates in fetal adrenals before and after incubation of minced tissue. He found that fetal adrenals synthesized a large amount of dehydroepiandrosterone sulfate and much smaller amounts of pregnenolone sulfate and 17-hydroxypregnenolone sulfate during a 4-hr incubation. During the same incubation, free pregnenolone and 17-hydroxypregnenolone decreased. This author was able to detect only a trace amount of dehydroepiandrosterone, suggesting very high steroid sulfotransferase activity towards this steroid.

Cooke and Taylor[39] determined the site of steroid sulfotransferase activity in human fetal adrenals. They observed that both zones in the adrenal of the previable fetus contain steroid sulfotransferase activity, but that the fetal zone exhibited a greater capacity for sulfating neutral steroids than did the adult zone.

Jaffe and Payne[40] investigated the relative capacity of human fetal adrenals and testes (from the same fetus) to sulfate neutral steroids. They observed that sulfation of the Δ^5-3β-hydroxysteroids, pregnenolone, and dehydroepiandrosterone was considerably greater in adrenals than in testes. Although there was extensive sulfation of testosterone in the adrenal, no sulfation of testosterone was observed in testes. This suggests that the steroid sulfotransferase of fetal testes may be specific for 3β-hydroxysteroids. The data also indicate that testosterone is maintained as a free steroid in fetal and adult testes.[28,29] This suggests that the free steroid or one of its unconjugated metabolites is essential for the androgen-mediated development of external genitalia in the human fetal male and for spermatogenesis in the adult male.

B. C_{21} and C_{19} Steroid Sulfates as Substrates for Biosynthetic Reactions

The direct interconversion of steroid sulfates has been demonstrated in vivo[2,3,41] and in vitro.[4,41-43] Baulieu et al.[2] administered [^{3}H]-androstenediol-[^{35}S]-(3-mono)-sulfate to two normal subjects and isolated double-labeled dehydroepiandrosterone sulfate from their urine. From the [^{3}H]/[^{35}S] ratios of the urinary dehydroepiandrosterone sulfate, they calculated that at least 90% of the oxidation of androstenediol in the male patient and 65% in the female patient in day 1 urine occurred via the direct pathway. In the same paper, these authors reported that the administration of androstenediol-3-sulfate to a male patient resulted in much greater amounts of dehydroepiandrosterone sulfate in the urine than was observed with androstenediol. This suggests that the steroid sulfate was a better substrate for the 17β-hydroxysteroid dehydrogenase than was the free steroid. A study by Payne and Mason[44] showed that a soluble extract of rat testes converted dehydroepiandrosterone sulfate to androstenediol-3-sulfate four times as rapidly as it converted the free steroid to androstenediol. The above reports suggest that steroid sulfates may be preferred over the free steroid as substrates by the steroid 17-oxidoreductases.

The direct metabolism of C_{21} steroid sulfates to other steroid sulfates has been reported by Calvin and co-workers both in vivo[2,41] and in vitro[41,42] in carcinomatous and hyperplastic adrenal tissue. They first[2] reported the direct conversion of pregnenolone sulfate to dehydroepiandrosterone sulfate in a female patient with adrenal cancer. In a subsequent study with a homogenate of hyperplastic adrenal tissue, Calvin and Lieberman[42] demonstrated the direct conversion of pregnenolone-^{3}H-sulfate-^{35}S to 17α-hydroxypregnenolone-^{3}H sulfate-^{35}S with a [^{3}H] to [^{35}S] ratio identical to the substrate. In another report from the same laboratory,[41] using double-labeled pregnenolone sulfate in vivo in a female patient with adrenal carcinoma and in vitro in homogenates of adrenal carcinomatous tissue, they demonstrated both 17-hydroxylation and reduction of C-20 of pregnenolone sulfate. The dehydroepiandrosterone sulfate and androstenediol-3-sulfate isolated from the patient's urine exhibited a [^{3}H]/[^{35}S] significantly lower than that in the injected compound. Conversion to double-labeled dehydroepiandrosterone sulfate was not observed in the in vitro incubation. From these results it was difficult to ascertain whether direct metabolism of pregnenolone sulfate to C_{19} steroid sulfates had taken place.

However, in a recent report from the same laboratory, Gasparini et al. demonstrated the direct conversion of a C_{21} steroid sulfate to two C_{19} steroid sulfates. Pregnenolone sulfate labeled with [^{3}H] and [^{35}S] was incubated with a microsomal preparation from boar testes. Two C_{19} steroid sulfates, dehydroepiandrosterone sulfate and 5,16-androstadien-3β-yl sulfate bearing the same [^{3}H]/[^{35}S] ratio as the substrate, were isolated and identified. This indicates that the boar testis has the capacity for metabolizing steroid sulfates directly to other steroid sulfates. The importance of this direct steroid sulfate pathway as compared to the free steroid pathway in porcine testes is not known at present.

Killinger and Solomon[45] studied the metabolism of [^{3}H] pregnenolone sulfate and [^{3}H] pregnenolone by homogenates of normal human adrenal tissue. They were unable to demonstrate any steroid sulfate products after incubation with pregnenolone sulfate, but identified the following products with pregnenolone as the substrate: pregnenolone sulfate, 17α-hydroxypregnenolone sulfate, and dehydroepiandrosterone sulfate. These results indicate that, in the normal human adrenal, pregnenolone serves as a substrate for further metabolism while pregnenolone sulfate is not metabolized. Pérez-Palacios et al.[43] studied the metabolism of [^{3}H] pregnenolone sulfate by homogenates of fetal adrenals of different gestational ages. They isolated 17α-hydroxypregnenolone sulfate, dehydroepiandrosterone sulfate, and 16α-hydroxydehydroepiandrosterone sulfate. They interpreted this as demonstrating a direct steroid sulfate pathway because of the absence of steroid sulfatase in second trimester fetal adrenals. However, in a later study, Payne and Jaffe[5] demonstrated that second trimester fetal adrenals do contain steroid sulfatase, and that it can cleave pregnenolone sulfate. The incubation media of Pérez-Palacios et al.[43] contained all the necessary cofactors for steroid sulfation. In addition, because very high steroid sulfotransferase activity has been reported in fetal adrenals, their study cannot be taken as proof of the presence of a direct steroid sulfate pathway in human fetal adrenal gland. Cook et al.[46] also studied pregnenolone and pregnenolone sulfate metabolism in adrenal glands of human previable fetuses. They reported very little metabolism of pregnenolone sulfate, although free pregnenolone was readily metabolized to pregnenolone sulfate, dehydroepiandrosterone sulfate, and dehydroepiandrosterone. The studies on human fetal adrenals indicate that if a direct steroid sulfate pathway exists, it is of minor significance. The major function of the fetal adrenal would appear to be the formation of steroid sulfates, which are transferred via cord blood to the placenta, where they are cleaved to free steroids and further metabolized to estrogens.

TABLE 4

Distribution of Mammalian Steroid Sulfatase

Tissue	Species	Steroid sulfatase specific activity[a] (pmol dehydroepiandrosterone sulfate hydrolyzed/ min/g wet tissue)
Liver	Rat[b]	84.7
	Guinea pig[c]	130.0
Kidney	Rat[b]	17.8
Seminal vescicle	Rat[b]	2.75
Levator ani	Rat[b]	2.08
Preputial	Rat[b]	0.90
Ventral prostate	Rat[b]	0.28
Ovary	Rat[b]	5.10
	Human[c]	220.00
Uterus	Rat[b]	3.20
Vagina	Rat[b]	0.54
Adrenal	Rat[c]	3560.00
	Guinea pig[c]	1060.00
	Human[c]	260.00
Testis	Bull[c]	180.00
	Ram[c]	180.00
	Boar	20.00
	Human[c]	260.00

[a] Measured with aliquots of homogenized tissue preparations.

[b] Data taken from Verde and Drucker.[55] Substrate concentration varied from 8 to 50 μM [^{3}H] dehydroepiandrosterone sulfate.

[c] Data taken from Burstein and Dorfman.[54] Substrate concentration was 100 μM [^{3}H] dehydroepiandrosterone sulfate.

C. Steroid Sulfatases

1. Tissue Distribution

A variety of steroid sulfates may serve as precursor pools for estrogen synthesis in the pregnant human female[6,47] and for androgen synthesis in the human male.[5,48,49] For their utilization as estrogen or androgen precursors, hydrolysis by steroid sulfatases is essential. Studies on the specificity of mammalian steroid sulfatases are limited. To date, only one such enzyme has been recognized by the Commission on Biochemical Nomenclature. This is sterol sulfate sulfohydrolase (E.C. 3.1.6.2.), which catalyzes the hydrolysis of dehydroepiandrosterone sulfate (17-oxoandrost-5-en-3β-yl sulfate). Other Δ^5-3β-yl sulfates are hydrolyzed by the same enzyme,[50,51] as discussed below. It is doubtful that estrone sulfate[52] or pregn-4-ene-C_{21}-yl sulfate[53] is hydrolyzed by the same enzyme as the Δ^5-3β-yl sulfates.

Steroid sulfatase activity has been found in many mammalian tissues. The distribution of this activity from various sources was reported by Burstein and Dorfman.[54] Verde and Drucker[55] reported on the distribution in a variety of tissues from male and female rats. These results are summarized in Table 4. French and Warren[56] reported that placenta is the richest source of human steroid sulfatase activity. Payne and coworkers demonstrated steroid sulfatase activity in rat[50] and human testes[51] and in human fetal gonads.[5,57] Studies by Kawano et al.[49] in human testes indicated that steroid sulfatase activity was found predominantly, if not solely, in seminiferous tubules.

TABLE 5

Steroid Sulfatase Activity in Human Fetal Tissue[a]

Estimated gestational age of fetus (days)	Tissue			
	Ovary	Testis	Adrenal	Liver
88	57.6		19.3	—
124	62.2		9.3	11.3
121	49.1		7.1	7.3
107		64.4	7.8	4.2
136		30.7	4.7	5.5
152		29.1	6.0	6.7

[a] Picomoles pregnenolone sulfate hydrolyzed per milligram protein during 1 hr incubation of tissue homogenates. Data taken from Payne and Jaffe.[45,57]

However, in rat testes,[58] steroid sulfatases are present in both testicular compartments, with interstitial tissue exhibiting the highest specific activity. Payne and Jaffe,[5,57] also demonstrated steroid sulfatase activity in second trimester human fetal livers and adrenals. Steroid sulfatase activity with pregnenolone sulfate was demonstrated in fetal livers and adrenal glands of both male and female fetuses, but the specific activity was considerably lower in fetal livers than in gonads (Table 5). Pulkkinen[59] was unable to detect steroid sulfatase activity in fetal rat livers using dehydroepiandrosterone sulfate or estrone sulfate. Using the same substrates, he also investigated several human fetal tissues. He observed no dehydroepiandrosterone sulfate hydrolysis in kidney, adrenal, liver, pancreas, or spleen. Only fetal liver exhibited steroid sulfatase activity, with estrone sulfate as the substrate.

Burstein and Westort[60] studied hepatic steroid sulfatase activity in rats during development. Low hepatic dehydroepiandrosterone sulfate hydrolysis was observed in the newborn. The enzyme activity, however, increased in livers of both males and females until 80 days. Higher levels were reached by males than females. Gill and Chen[61] observed steroid sulfatase activity in both the dorsal and the ventral prostate, the coagulating gland, and seminal vesicles from male rats. Comparison of a number of steroids showed that all of the above tissues exhibited the highest activity with estrone sulfate as substrate, followed by pregnenolone sulfate and dehydroepiandrosterone sulfate. No enzyme activity could be demonstrated with either androsterone sulfate or testosterone sulfate. From these studies, it is not possible to ascertain whether the same enzyme catalyzed the hydrolysis of both the Δ^5-3β-yl sulfate and the estra 1,3,5,(10) triene-3-yl sulfate. Steroid sulfatase activity with dehydroepiandrosterone sulfate was reported by Farnsworth[62] in human prostate glands. He showed that benign epithelial hyperplasia exhibited a higher steroid sulfatase activity per milligram of tissue than did prostate glands with stromal hyperplasia or adenocarcinoma.

2. Cellular Localization

Mammalian steroid sulfatases are found in microsomal fractions of the cell.[53,54] Studies in rat liver demonstrated that this activity was restricted to the rough endoplasmic reticulum during early development, and that it subsequently became equally distributed in rough and smooth endoplasmic reticulum.[60] Investigations on specificity and other characteristics of steroid sulfatases have been hindered by the fact that steroid sulfatases are membrane bound. Various solubilization techniques have been unsuccessful.[49,63] French and Warren[52] subjected a microsomal preparation from human placenta to a variety of different treatments and concluded that there are two distinct steroid sulfatases, one that hydrolyzes dehydroepiandrosterone sulfate and another that hydrolyzes estrone sulfate. They suggested, on the basis of sensitivity to ribonuclease and to lipase, that steroid sulfatases are bound to a particle containing lipid and RNA and that enzymic activity is dependent on structural integrity of this particle.

3. Testicular Steroid Sulfatase

Numerous studies using a variety of tissues as source of enzyme indicate that steroid sulfatases have the properties of regulatory enzymes and might control the amount of free steroids available for further metabolism to active hormone. Payne et al.,[50] using a partially purified particulate preparation of rat testes, demonstrated that dehydroepiandrosterone sulfate and androstenediol-3-sulfate were hydrolyzed by the same enzyme. The K_ms for dehydroepiandrosterone sulfate and androstenediol-3-sulfate were 2 μM and 0.9 μM, respectively. Inhibition studies with 15 C_{18} and C_{19} free steroids demonstrated that 5α-androstane-3β,17β-diol was the most potent inhibitor tested. The most important structural features for maximum inhibition were a saturated A ring and a 3β and 17β hydroxyl group. K_i values for three of the steroids tested were 1.7 μM for 5α-androstane-3β,17β-diol, 3.3 μM for 5α-androstane-3α,17β-diol, and 11.8 μM for testosterone. Kinetic data (Figure 1A and B) indicate that 5α reduced C_{19} steroids inhibit dehydroepiandrosterone sulfate hydrolysis competively, but at a site distinct from the catalytic site. Figure 1A is a Lineweaver-Burk plot of data obtained when increasing concentrations of dehydroepiandrosterone sulfate were incubated with four different fixed concentrations of 5α-androstane-3α,17β-diol. Incubation of dehydroepiandrosterone sulfate in the presence of 5α-androstane-3α,17β-diol resulted in an increase in K_m values for dehydroepiandrosterone sulfate, but no change in V_{max}. A replot of the data from Figure 1A showed that when 1/v was plotted as a function of increasing concentrations of free steroid, hyperbolic curves resulted (Figure 1B). This is characteristic of partial competitive inhibition, indicating that the free steroid inhibitor combines with the enzyme at a different site than the substrate. Similar studies carried out by Payne,[52] with microsomal preparations from human testes demonstrated that human testes steroid sulfatase exhibits the same characteristics (Figure 2). In these studies, it was observed that pregnenolone sulfate exhibits a lower K_m than either androstenediol-3-sulfate or dehydroepiandrosterone sulfate. The respective K_m values were 0.73, 3.13, and 3.85 μM. Kinetic studies showed that all three substrates were hydrolyzed by the same enzyme.[51] Several C_{21} steroids were tested for inhibition of steroid sulfatase activity. Pregnen-3β,21-diol-20-one and 5-pregnen-3β,20α-diol were the most potent inhibitors.

Notation and Ungar[64] reported that 5-pregnen-3β,20α-diol was the most effective inhibitor of pregnenolone sulfate hydrolysis in rat testicular homogenates. Townsley et al.[65] have shown that the 5-pregnen-3β,21-diol-20-one was the most potent inhibitor of the placental steroid sulfatase.

Among the C_{19} steroids tested, 5α-reduced steroids appeared to be more potent inhibitors than Δ^4 or Δ^5 steroids. Studies with human testicular preparations showed that the 5α-androstane-3α,17β-diol was a more effective inhibitor than the 5α-androstane-3β,17β-diol, as had been found for the enzyme from rat testes. The kinetics of inhibition of the human testicular steroid sulfatase by the free steroids are consistent with partial competitive inhibition (Figure 2), as also demonstrated for the enzyme from rat testes. Another property characteristic of regulatory enzymes, substrate inhibition at high substrate concentration, was observed in the studies with the enzyme preparations from both rat and human testes.[50,51]

The C_{19} and C_{21} steroids that were the most potent inhibitors for the steroid sulfatase in rat and human testes have been identified as metabolic products in studies with testicular preparation from these sources. Payne and Jaffe[66] reported the isolation of

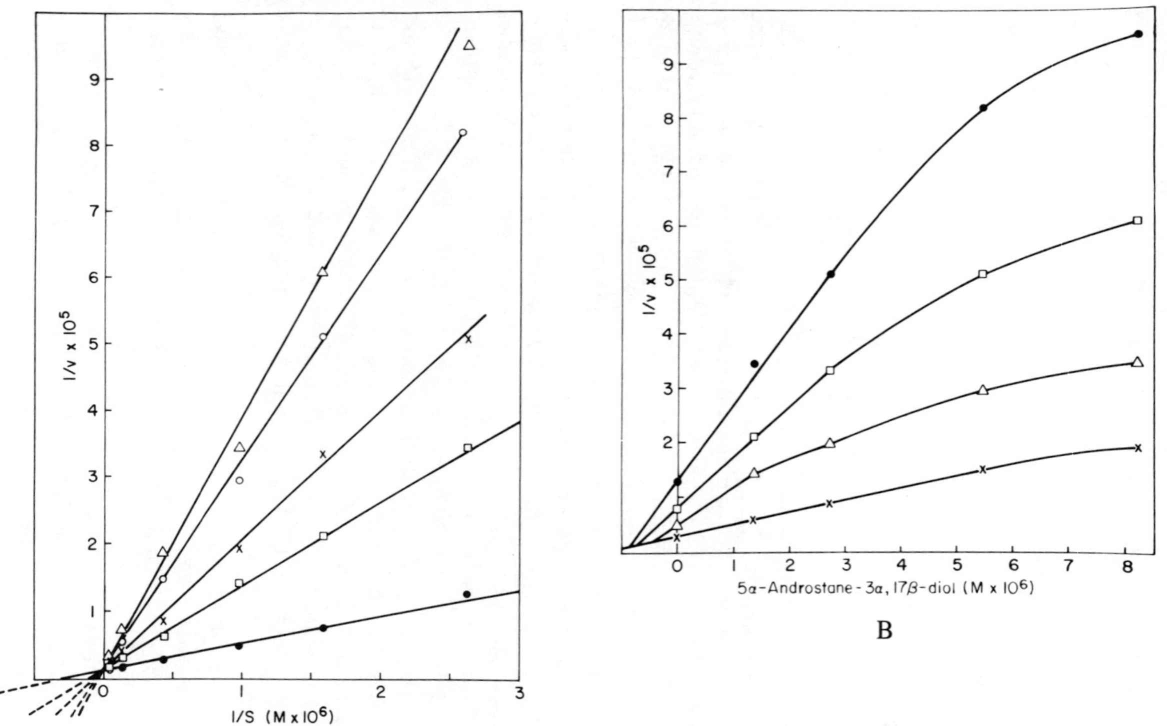

FIGURE 1. Inhibition of hydrolysis of dehydroepiandrosterone sulfate by 5α-androstane-3α, 17β-diol. Rat testicular microsomal preparations (0.46 mg protein) were incubated with indicated amount of substrate and inhibitor for 15 min at 37°C. A. Double reciprocal plots of velocity vs. concentration of dehydroepiandrosterone sulfate at the following fixed concentrations of 5α-androstane-3α, 17β-diol; O, •—•; 6.94 μM, □—□; 13.9 μM, x—x; 27.8 μM, o—o; 41.7 μM, Δ—Δ. B. Replot of data from A. The reciprocal of the velocity as a function of increasing concentrations of the inhibitor, 5α-androstanediol, with the following fixed concentrations of the substrate, dehydroepiandrosterone: 0.37 μM, •—•; 0.59, μM, □—□; 1 μM, Δ—Δ; 2.3 μM, x—x. (From Payne, A. H., Mason, M., and Jaffe, R. B., *Steroids*, 14, 685, 1969. With permission.)

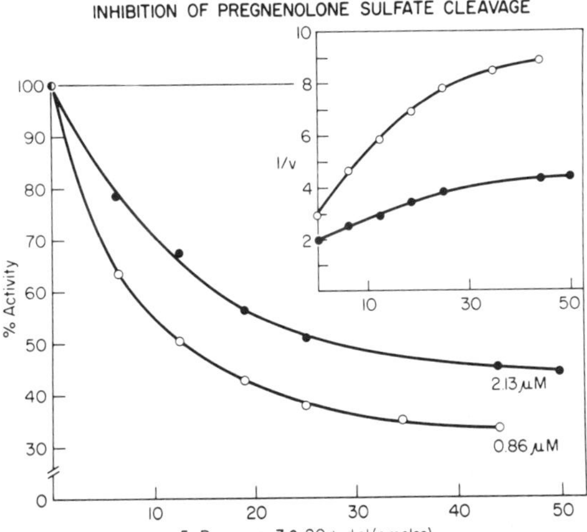

FIGURE 2. Inhibition of hydrolysis of pregnenolone sulfate by 5-pregnene-3β,20α-diol. Human testicular microsomal preparations (0.5 mg protein) were incubated with indicated amount pregnenolone sulfate in the presence of increasing concentrations of 5-pregnen-3β,20α-diol for 20 min at 37°C. Inset represents reciprocal of the velocity as a function of increasing concentrations of the inhibitor 5-pregnene-3β,20α-diol. (From Payne, A. H., *Biochim. Biophys. Acta,* 258, 473, 1972. With permission.)

5α-androstane-3β,17β-diol and 5α-androstane-3α,17β-diol in incubations of androstenediol-3-sulfate, with a microsomal preparation from rat testes. They observed that 5α-androstane-3β,17β-diol, which had been shown to be the most potent C_{19} steroid inhibitor of the sulfatase activity,[50] was the most abundant free steroid identified other than testosterone and androstenedione. In studies with isolated seminiferous tubules and minced tissue from human testes, Kawano et al.[50] isolated 5-pregnene-3β,20α-diol in incubations with pregnenolone sulfate as the substrate. They demonstrated that steroid sulfatase and 20α reductase, the enzyme that converts pregnenolone to 5-pregnene-3β,20α-diol, are localized predominantly in seminiferous tubules of the human testes. Payne and co-workers[49,67] and Rivarola[68] reported that the major concentration of 5α reductase, the enzyme essential for the production of 5α-androstane-3α,17β-diol, is also found predominantly in seminiferous tubules of human testes.

Presently, no data is available on the testicular site of steroid sulfate formation in human testes. However, as reported earlier in this chapter, steroid sulfotransferase activity in rat testes has been found only in seminiferous tubules. If this is also the case in human testes, the high concentration of steroid sulfates found in these organs would be synthesized in the same compartment in which they might serve as precursors of free steroid. In addition, if pregnenolone or dehydroepiandrosterone is sulfated in seminiferous tubules, no mechanism for transport of these polar steroid compounds across the plasma membrane and across the blood-testis barrier need exist in order for these compounds to serve as potential precursors of active androgens. It has been demonstrated by several investigators that pregnenolone can cross the blood-testis barrier. Sulfation of free steroids takes place in cytoplasm. It is possible that a specific binding protein exists in seminiferous tubules similar to that which has been described for sulfate esters of glucocorticoids in rat liver cytoplasm.[69-71] This proposed binding protein might concentrate steroid sulfates in seminiferous tubules and transport them to the endoplasmic reticulum for hydrolysis by steroid sulfatase. Then they could be further metabolized to testosterone and dihydrotestosterone. This pool of steroid sulfates might serve as an auxiliary source of androgens in seminiferous tubules where testosterone or dihydrotestosterone is essential for maintaining normal spermatogenesis. According to the properties of steroid sulfatases described above, it might be postulated that free steroids, as they accumulate in seminiferous tubules, could be acting as inhibitors of the steroid sulfatase.

4. Placental Steroid Sulfatase

Placental steroid sulfatase appears to determine the amount of maternal estrogens found in serum and urine during pregnancy. In late pregnancy, approximately 90% of estriol and 60% of estrone and estradiol are derived from dehydroepiandrosterone sulfate and 16α-hydroxydehydroepiandrosterone sulfate, originating in the fetus.[48] The placental enzymes, steroid sulfatase, 3β-hydroxysteroid dehydrogenase-isomerase, and aromatase, bring about conversion of fetal steroid sulfates to estrogens. The importance of placental steroid sulfatase in determining maternal serum and urinary estriol levels has been demonstrated in several cases of placental steroid sulfatase deficiency.[72-75] In patients with placental steroid sulfatase deficiency, estriol excretion is approximately 5% of that found in normal pregnant women.[73,75]

Evidence presented by Townsley et al.[65,76] suggests that in the placenta, the concentration of endogenous free steroids may regulate placental estrogen production from conjugated precursors, due to the inhibitory effect of free steroids on steroid sulfatase. These authors examined the effect of estrogens, intermediates of estrogen synthesis, and various steroids present in cord blood on steroid sulfatase activity in the 10,000 × g supernatant fraction of placental homogenates. In addition, they examined the effect of pregnenolone sulfate on the hydrolysis of dehydroepiandrosterone sulfate. The results obtained were analogous to those found with the testicular steroid sulfatase. Pregnenolone sulfate was a better substrate than dehydroepiandrosterone sulfate for the placenta steroid sulfatase. C_{21} sterois were more active inhibitors than C_{19} steroids. Inhibitory activity of unconjugated steroids was favored by planar Δ^5 — or 5α structures unsubstituted except for oxygen functions at C-3 and C-20. It was also demonstrated that there is a cumulative effect of unconjugated steroids on the steroid sulfatase activity; thus, the total concentration of endogenous free steroids in the placenta may contribute to the regulation of estrogen synthesis from dehydroepiandrosterone sulfate and 16α-hydroxydehydroepiandrosterone sulfate.

5. Prostatic Steroid Sulfatase

Steroids derived from dehydroepiandrosterone sulfate by the action of steroid sulfatase may be of importance in the maintenance of the prostate gland. It has been reported[77] that the ventral prostate developed in immature, castrate rats, but not in adrenalectomized, immature, castrate rats. Lostroh and Li[78] reported that corticosterone, cortisone, and hydrocortisone are without effect on the weight of ventral prostates in hypophysectomized, castrated, adult rats. Since it has been reported that both human and rat prostates contain steroid sulfatase that can readily hydrolyze dehydroepiandrosterone sulfate, it seems possible that the steroid sulfatase activity of the prostate gland may, in part, determine the amount of free androgens present for the maintenance of prostatic weight. As noted earlier in this chapter, Farnsworth[63] reported that steroid sulfatase activity was considerably greater in glands exhibiting epithelial hyperplasia than in those with stromal hyperplasia or adenocarcinoma. He also reported that testosterone, dihydrotestosterone, and 5α-androstane-3α,17β-diol at relatively low concentrations are effective inhibitors of the prostatic steroid sulfatase. Diethylstilbestrol, which is used for treatment of prostatic carcinoma, was also found to be a very effective inhibitor of steroid sulfatase activity when tested at low concentrations in vitro.

TABLE 6

Effect of Gonadotropin Treatment on Rat Testicular Steroid Sulfatase[a]

Treatment	Interstitial tissue	Seminiferous tubules	Whole testes
Saline	119 ± 7[b]	21.0 ± 1.6	31.7 ± 2.8
LH[c]	135 ± 10	21.6 ± 2.0	32.2 ± 1.7
FSH[d]	161 ± 9[e]	24.4 ± 3.2	30.4 ± 1.5

[a] Picomoles dehydroepiandrosterone sulfate hydrolyzed per minute per milligram microsomal protein.
[b] Mean ± S.E., n = 7.
[c] 20 gmg LH (NIH-LH-S19) was administered twice per day for 6 days.
[d] 50 μg FSH (NIH-FSH-S11) was administered twice per day for 6 days.
[e] p = < 0.01 compared to saline injected controls. Data from Georgopoulos and Payne.[80]

6. Pituitary Control of Steroid Sulfatase

Notation and Ungar[79] investigated the effect of administration of hCG to intact rats for 3 and for 10 days and of hypophysectomy on the cleavage of pregnenolone sulfate by microsomal fractions from rat testes. No effect on the testicular steroid sulfatase activity was observed in either the hypophysectomized rats 3 days after surgery or in the hCG-treated rats. Payne and Kelch[58] investigated steroid sulfatase activity in homogenates from isolated testicular compartments as well as from whole testicular tissue 2 weeks following hypophysectomy. Hypophysectomy had no effect on steroid sulfatase activity in seminiferous tubules, but it significantly decreased the specific activity in interstitial tissue. Because interstitial tissue in rat testes comprises a small percentage of total testicular tissue, no effect of hypophysectomy could be discerned in the steroid sulfatase activity from whole testicular tissue. In subsequent studies, Georgopoulos and Payne[80] investigated the effect of in vivo treatment with follicle stimulating hormone (FSH) or luteinizing hormone (LH) on steroid sulfatase activity in isolated testicular compartments from adult rats. Adult male rats received two subcutaneous injections per day of either FSH, at 50 μg per injection, or LH, at 20 μg per injection, for 6 days. Control animals were injected with saline. Rats were killed on day 7 and the testes removed and processed. Steroid sulfatase activity was determined in microsomal fractions of both interstitial tissue and seminiferous tubules by measuring the amount of free steroid produced from dehydroepiandrosterone sulfate. The results of this study are presented in Table 6. As reported earlier in this chapter, steroid sulfatase is found in both testicular compartments of the rat with the higher specific activity demonstrated in interstitial tissue. As shown in Table 6, neither FSH nor LH treatment had an effect on steroid sulfatase activity in seminiferous tubules. This is in agreement with the earlier observation by Payne and Kelch[58] that hypophysectomy had no effect on steroid sulfatase in seminiferous tubules. Steroid sulfatase activity in interstitial tissue, however, was significantly increased in FSH treated rats as compared to saline-injected controls ($p < 0.01$). LH treatment had no significant effect on interstitial tissue steroid sulfatase activity. These results indicate that interstitial tissue steroid sulfatase activity may be regulated by FSH. Because no effect was observed with LH, the FSH effect on interstitial tissue steroid sulfatase cannot be due to LH contamination. Whether this is a direct effect of FSH on interstitial tissue or an indirect effect mediated by the seminiferous tubules cannot be ascertained from these experiments.

Evidence that steroid sulfatase might be regulated by ACTH in adrenal glands has been presented by Dominguez et al.[81] They reported that treatment of adult female rats for 3 weeks with daily intravenous injections of ACTH resulted in an increase of steroid sulfatase activity in adrenal homogenates with either pregnenolone sulfate or dehydroepiandrosterone sulfate as the substrate, when compared to saline-injected controls.

III. GLUCOCORTICOID SULFATES

A. Corticosteroid Sulfates in Biological Fluids and in Steroid Metabolism

Glucocorticoid sulfates are present in large amounts in the human feto-placental circulation and in the urine of infants.[82,83] They are also found in the blood and urine of adult subjects.[84-86] In addition, corticosteroid sulfates are the predominant forms of these steroids in rat bile.[87,88]

In rats, large amounts of anionic metabolites,[89,90] sulfated glucocorticoid metabolites,[91-93] were found in the liver soon after administration of physiological amounts of ^{3}H-cortisol. Recently, similar results have been reported with corticosterone by Carlstedt-Duke and Gustafsson.[94] There is also evidence that corticosteroid sulfates could be involved in the control of some facets of the metabolism of related steroids, via the action of specific steroid hydroxylases and reductases that prefer them to the analogous unconjugated steroids. Such a role for corticosteroid sulfates may be similar to the role of estrogen sulfates in estrogen metabolism suggested by Miyazaki et al.[95] and supported by the studies of Brooks and Horn.[96]

Investigations on steroid hydroxylases and reductases that prefer corticosteroid sulfates to the analogous free steroid include studies by Gustafsson and Ingleman-Sundberg, on the microsomal 16β-hydroxylase of female rat livers[97,98] and on the 3β-, 5α-, and 5β-steroid reductases in livers of rats of both sexes.[99] In the case of the reductases, corticosterone-21-sulfate and deoxycorticosterone-21-sulfate were as good as or better substrates than the analogous unconjugated steroids. The 20β-reductase, present only in males, used only the sulfated steroids as substrates. These studies indicate that corticosteroid sulfates may play an important role as substrates for further metabolism in the liver.

B. Corticosteroid Sulfatases

An understanding of the biochemical roles of the corticosteroid sulfates requires the thorough elucidation of their biosynthesis and degradation. Presently, very little information exists concerning these areas of corticosteroid metabolism. Detailed examination of the hydrolysis of corticosteroid sulfates by steroid sulfatases is almost nonexistent. Studies by those few researchers who touched on the area agree generally with the report by Hall and Giroud,[53] which showed that placental homogenates contain little ability to hydrolyze corticosteroid sulfates, compared to dehydroepiandrosterone sulfate.

C. Corticosteroid Sulfotransferases

There are few definitive reports concerning the sulfation of corticosteroids. Most studies are limited to preliminary examination of crude cytosol preparations.[100-112] In some cases, data reported by different laboratories are in conflict. Information about hepatic enzyme levels or organ and subcellular distribution of corticosteroid sulfotransferase activity is also limited. Many of the studies on corticosteroid sulfotransferase activity used ATP as a coenzyme instead of 3′-phosphoadenosine-5′-phosphosulfate (PAPS).

The use of ATP as the coenzyme when measuring sulfotransferase activity is permissible only if the enzymes involved in the synthesis of PAPS[113,114] are not rate limiting in the tissue samples studied. However, in general, no attempts have been made to ascertain this fact for the corticosteroid sulfotransferases. One researcher[111] who used ATP based his action on a report attributed to Wengle[115] that "cytoplasmic dehydroepiandrosterone sulfation in livers from rats of different ages and both sexes was never limited by the ability of cytosols to accumulate PAPS from added ATP." Studies by Singer et al.[91,116] have shown that great variations occur in the determination of corticosteroid sulfation, unless PAPS is used as the coenzyme.

A more adequate understanding of the biochemical roles of the corticosteroid sulfates requires a more complete understanding of the enzymes that produce them. Questions to be answered regarding these enzymes include:

1. Are there specific corticosteroid sulfotransferases?
2. What are their properties?
3. How is their production controlled?
4. Can they be related to any significant effects of the corticosteroids?

The remainder of this chapter is an attempt (Singer and co-workers)[91,116-119] to answer these four questions as they pertain to the corticosteroid sulfotransferases of rat liver.

1. Development of a Corticosteroid Sulfotransferase Assay

The basic assay method developed by Singer et al.[116] used the ^{3}H-hormone and unlabeled PAPS. After incubation, the unreacted substrate was extracted into organic solvent and discarded. The extent of sulfation was determined by counting the aqueous residue, which contained the steroid sulfate. This assay was simpler than the assay used by Adams and Poulos[120] for the estrogen sulfotransferase (originally described by Wengle)[121] that used ^{35}S-PAPS and nonradioactive steroid. It was also less prone to error than the relatively unspecific methylene blue assay method of Roy.[122]

The first consideration was the choice of the substrate to be used. Three adrenal corticosteroids were tested: cortisol, corticosterone, and deoxycorticosterone. Quantitatively, deoxycorticosterone was the poorest substrate. Corticosterone was a somewhat better substrate than cortisol. However, as shown in Figure 3, both corticosterone and deoxycorticosterone exhibited marked substrate inhibition, which complicated the assay. This substrate inhibition required the use of a cumbersome "mosaic" assay with several steroid levels, each studied with varying amounts of enzyme sample, for reliable results.[91,119] Cortisol sulfotransferase activity exhibited simple Michaelis-Menten kinetics. Therefore, cortisol was chosen for the routine study of the corticosteroid sulfotransferase activity of rat liver. Of the organs studied, liver is the richest source of the enzyme activity.[91] The position of sulfation of all three corticosteroids by the rat liver enzyme activity was at carbon-21.[91,116-119]

2. Fractionation of the Cortisol Sulfotransferase Activity of Livers from Male and Female Rats

Sex differences in the steroid sulfotransferase activity of rat liver have been described for several hormonal steroids.[109,111,112,115] Therefore, it was not surprising that a sex difference in cortisol sulfotransferase activity was found.[116] Female rats had six to eight times the hepatic enzyme activity found in males. Interestingly, the enzyme activity in the two sexes exhibited very different substrate preferences. This suggested that the steroid sulfotransferases in male livers were different from those in female livers. Fractionation of cytosol samples from females on columns of DEAE Sephadex A-50 (Figure 4A) showed that they contained three enzymes that sulfated cortisol. These enzymes, present in roughly equal proportions, were named sulfotransferases I, II,

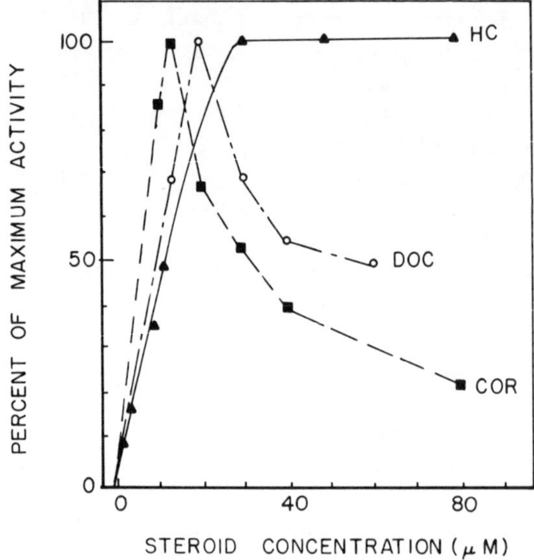

FIGURE 3. Sulfation of cortisol (HC), corticosterone (COR), and 11-deoxycorticosterone (DOC) by the corticosteroid sulfotransferase activity of liver cytosol from male rats. The enzyme assay is described in Reference 116. The data are given as percent of maximum activity. The steroid concentration in the reaction mixtures is varied as indicated. Enzyme assays were carried out with PAPS as coenzyme.

and III (STI, STII, and STIII). Cytosols from male rats (Figure 4B) contained only two sulfotransferases. The major enzyme, comprising up to 90% of the recovered activity, eluted in a similar way to that of STIII from females. It was tentatively named "STIII". The minor enzyme was named "STII" for similar reasons. It comprised 10 to 20% of the recovered enzyme activity. Figure 4A also shows the importance of assaying the eluted column fractions with PAPS. If ATP is used as the pseudocoenzyme, only STIII is observed. It appears probable that unlike the other sulfotransferases, STIII cochromatographs fortuitously with the enzymes of PAPS synthesis.

The three sulfotransferases exhibit very different substrate preferences. STII either elutes concurrently with, or is the same enzyme as, the dehydroepiandrosterone sulfotransferase reported by Ryan and Carroll.[10] None of the three steroid sulfotransferases described appears to be identical to the "specific" estrogen sulfotransferases of Adams and Poulos,[120] since most of the estrogen sulfotransferase activity of cytosols is lost after DEAE Sephadex A-50 chromatography.[116] However, STI, STII, and STIII all possess the ability to sulfate significant amounts of estradiol-17β. An understanding of the absolute substrate specificity of each of the three enzymes must await their successful purification to homogeneity.

3. Development of the Cortisol Sulfotransferase Activity of Livers from Growing Rats

The initial examination of the development of cortisol sulfotransferase activity (Figure 5) showed that it was very low in rats of either sex 2 days after birth. After this, the enzyme activity developed in close parallel in both sexes until day 28 to 30 after birth. At this time, the activity in males began to drop and that in females began to rise rapidly. By approximately 2 months after birth, essentially "adult" enzyme patterns were observed in both sexes. At this time, livers from females contained six to eight times the activity found in those from males.

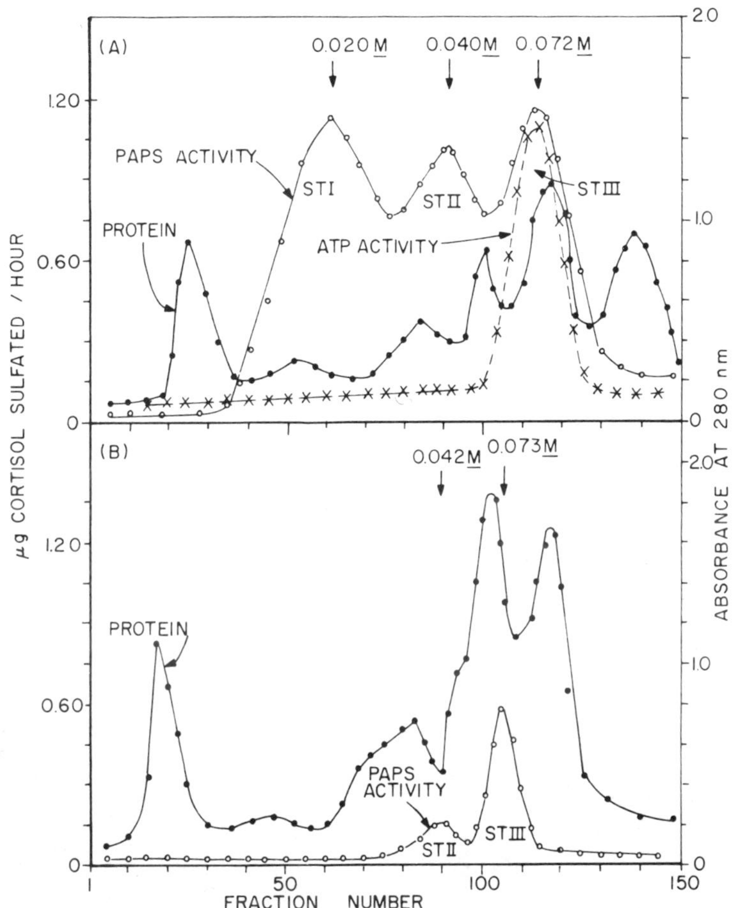

FIGURE 4. Chromatography of the cortisol sulfotransferase activity of liver cytosols from male and female rats on DEAE Sephadex A-50 columns. A 6-mℓ sample of 50% cytosol from (A) female or (B) male rats was loaded on a 3 × 50 cm column of DEAE Sephadex A-50. Columns were eluted with linear gradients consisting of 600 mℓ each of 0.05 *M* tris-0.25 *M* sucrose-0.003 *M* mercaptoethanol (TSM), pH 7.5, and TSM containing 0.30 *M* KCl; 6-mℓ fractions were collected. Protein was estimated from the absorbance of 280 nm. Aliquots, 0.50 mℓ, of the indicated fractions were assayed for cortisol sulfotransferase activity with PAPS (o—o) or ATP (x—x). Enzyme activity is given as micrograms of cortisol sulfated per hr per aliquot. When assayed with PAPS, three peaks of cortisol sulfotransferase activity were observed in cytosol samples from females. They were named STI, STII, and STIII in the order of their elution from columns. The molar concentrations represent peak KCl concentrations used for elution of the individual enzymes. Preparations from males exhibit two enzyme peaks, tentatively assigned as STII and STIII. Enzyme recoveries were 85 and 89% for the preparations from male and female rats, respectively. (From Singer, S. S., Giera, D., Johnson, J., and Sylvester, S., *Endocrinology*, 98, 963, 1976. With permission.)

When cortisol sulfotransferase activity was assayed with ATP and the activity compared to that with PAPS, significant differences were found in the activity ratio (ATP/PAPS). These occurred at different stages of growth both within a sex and between sexes. For example, the ATP/PAPS in females was 0.35 to 0.45 until 30 days after birth. Then, within the next month, it rose to a maximum of 0.60 to 0.72. In males, a similar trend was observed; however, the ATP/PAPS in adult animals (e.g., over 2 months old) was 0.90 to 0.96. These differences in the ATP/PAPS point out the weak-

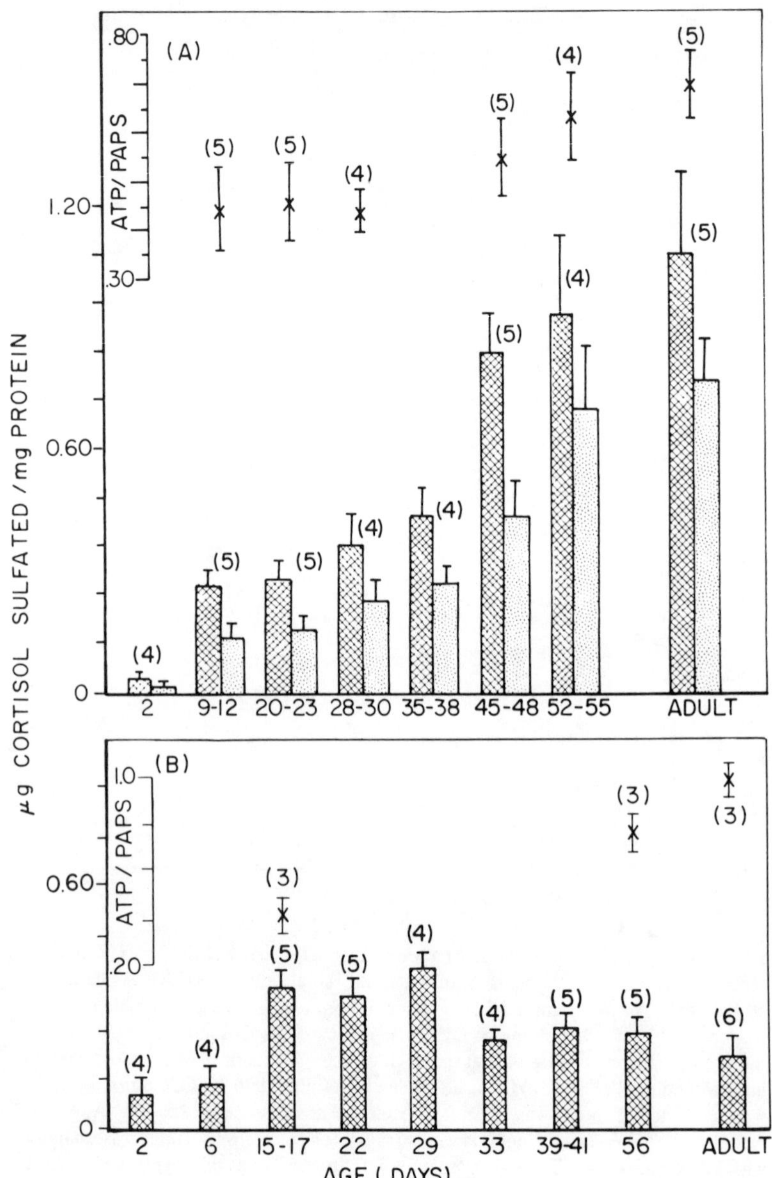

FIGURE 5. Development of cortisol sulfotransferase activity in livers from male and female rats. Cytosols were prepared from (A) female or (B) male rats of the indicated ages. Cortisol sulfotransferase activity was measured with PAPS, cross-hatched bars, or ATP, stippled bars. The enzyme activity is given as micrograms of cortisol sulfated per milligram of cytosol protein (measured by the biuret assay) in 60 min. The inserts represent the ratio of activity with ATP to that with PAPS. The ATP data are not shown in B because only three experiments were carried out with both ATP and PAPS at each age indicated. The vertical lines are the standard deviations. The numbers in parentheses are the number of experimental points. (From Singer, S. S., Giera, D., Johnson, J., and Sylvester, S., *Endocrinology*, 98, 963, 1976. With permission.)

ness of the use of ATP even in assaying cortisol sulfotransferase activity in cytosols. The variation of the total activity assayed with PAPS and the activity observed with ATP might account for the differences between our developmental pattern and the developmental pattern reported by Carlstedt-Duke and Gustafsson[111] for the deoxycorticosterone sulfating activity of rat liver.

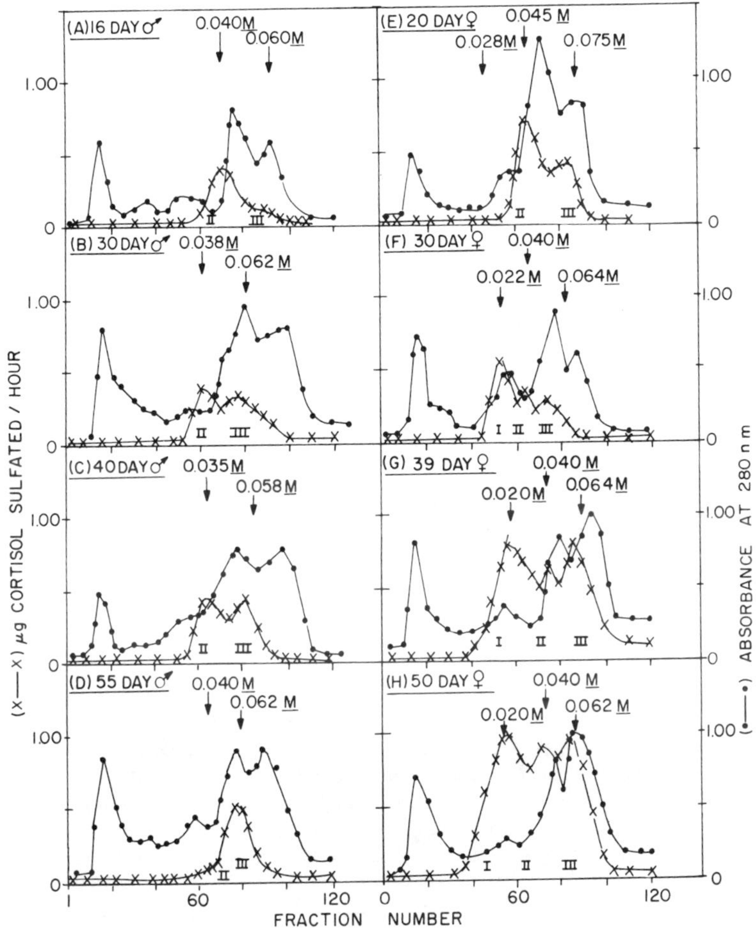

FIGURE 6. DEAE Sephadex A-50 chromatography of cytoplasmic cortisol sulfotransferase activity from livers of male and female rats of different ages. For these representative experiments, paired columns were always run. One column was loaded with 2.5 m*l* of cytosol from an adult rat, the other with an equivalent amount of cytosol from a rat of the same sex and indicated age. Enzyme activity with PAPS (x—x); protein (o—o). Assignment of the presence of STI, STII or STIII was made by comparison with adult columns. These are not shown. All conditions are the same as in Figure 4 except that 2 × 50 cm columns were used and the gradients consisted of 300 m*l* each of TSM and TSM-KCl; 3-m*l* fractions were collected. (From Singer, S. S., Giera, D., Johnson, J., and Sylvester, S., *Endocrinology,* 98, 963, 1976. With permission.)

Because three hepatic enzymes sulfate cortisol, total activity data could mask important differences in the relative amounts of the individual sulfotransferases present in rats of different ages. Therefore, development of the individual enzymes was examined (Figure 6). In immature rats of both sexes (Figures 6A and E), the major cortisol sulfotransferase was STII. By 30 days after birth (Figure 6B), STIII levels began to rise in males. After this time, the rise of STIII continued and STII dropped until, by 2 months after birth (Figure 6D), STIII accounted for 80 to 90% of the cortisol sulfotransferase activity in males. No measurable amounts of STI were observed in males at any time. In females, STI was the major enzyme in liver by day 30 (Figure 6F). After this, the production of all three enzymes increased (Figures 6G and H) until roughly adult enzyme patterns containing similar amounts of STI, STII, and STIII were obtained by day 50.

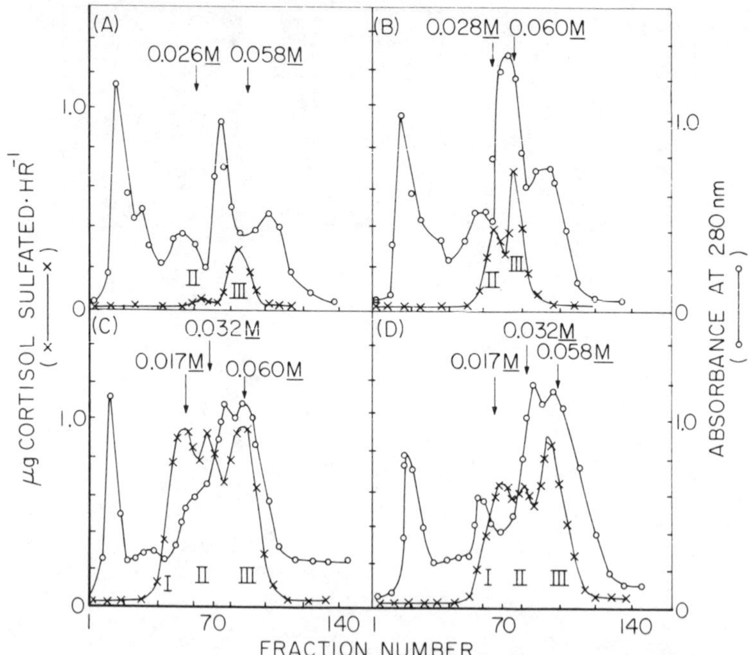

FIGURE 7. Fractionation of the cytoplasmic cortisol sulfotransferase activity of liver from intact and castrated rats on DEAE Sephadex A-50 columns. Samples of 3 mℓ of cytosol from (A) intact or (B) castrated males and (C) intact or (D) ovariectomized females were loaded on 2 × 50 cm columns of DEAE Sephadex A-50. The columns were treated as described in Figure 6. The columns described in A and B and those in C and D were each run as simultaneous paired experiments. The recoveries of enzyme activity were 79 to 88% of the original activity loaded. The data are from one of three similar experiments. (From Singer, S. S. and Sylvester, S., *Endocrinology*, 99, 1346, 1976. With permission.)

The data are consistent with the concept that the production of all three sulfotransferases is controlled by the gonads, with STI and STII being most sensitive to these endocrine effects. It would appear that the ovaries might stimulate the production of maximal amounts of cortisol sulfotransferase activity and that the testes might inhibit production of the enzymes. It should be mentioned here that recent observations indicate that STI, STII, and STIII may be considered as glucocorticoid sulfotransferases. We have evidence that these enzymes not only sulfate cortisol, but are also responsible for sulfation of corticosterone.[119]

D. Endocrine Control of the Production of Hepatic Cortisol Sulfotransferase Activity

The control of the production of the sulfotransferases by gonadal ablation and/or sex hormone replacement therapy was examined by Singer and Sylvester.[118] A comparison of intact rats and rats castrated 28 days after birth and sacrificed 45 to 88 days later showed that cortisol sulfotransferase activity levels were more than twice as high in the castrated males than in the control rats. Ovariectomy decreased the enzyme activity in females by only 25 to 35%. Within each sex, liver and body weights of intact and castrated animals did not markedly differ.

Fractionation of cytosols from male rats on DEAE Sephadex A-50 columns showed that the elevated enzyme activity in castrates was due largely to increased STII levels (compare Figures 7A and B). Similar studies with females (Figures 7C and D) showed that the decreased enzyme activity in cytosols from ovariectomized rats was due to the diminuition of STI and STII. STIII levels were not affected significantly.

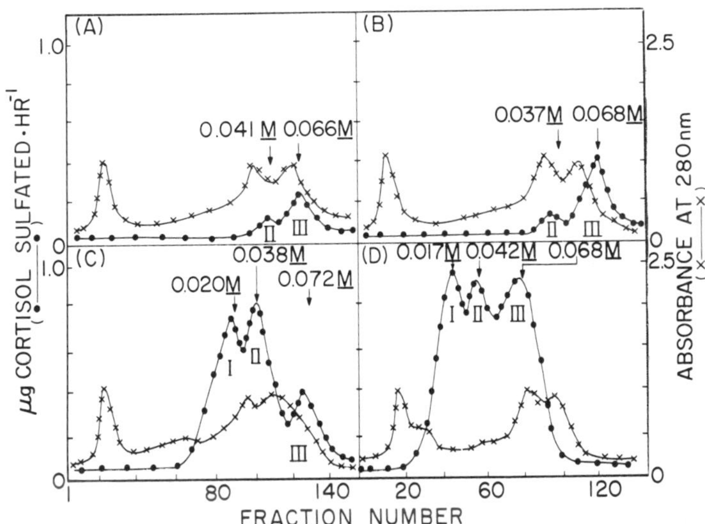

FIGURE 8. Fractionation of cortisol sulfotransferases of cytosols from male rats subjected to various endocrine treatments. Samples of 3 mℓ cytosol from intact males given (A) oil, (B) testosterone, (C) estradiol-17β and (D) castrated males given estradiol-17β were fractionated on 2 × 50 cm DEAE Sephadex A-50 columns. Enzyme activity (•—•); protein (x—x). Conditions are as described in Figure 6. The data are from one of three similar experiments. (From Singer, S. S. and Sylvester, S., *Endocrinology*, 99, 1346, 1976. With permission.)

It appears from the studies with castrated male and female rats that the three sulfotransferases are differentially controlled by the gonadal hormones. To determine the nature of the gonadal factors involved in control of hepatic cortisol sulfotransferase activity, the effects of chronic daily administration of 1 mg of testosterone or 200 μg of estradiol-17β to intact rats were examined. In males, 50 to 70 days of testosterone injection elevated the cortisol sulfotransferase activity by 70 to 80%. Similar estradiol administration to males elevated the enzyme activity nine- to tenfold, resulting in enzyme levels comparable to those in females. Larger doses of the estrogen were lethal.

Similar experiments were carried out with intact females. These showed that estradiol administration for 2 to 3 months did not increase the cortisol sulfotransferase activity above the levels found in untreated controls. Testosterone administration for 50 to 70 days resulted in a 50 to 70% decrease of the enzyme activity. These data support a role for sex hormones as the active agents in gonadal control of hepatic cortisol sulfotransferase activity. Studies with castrated rats of both sexes gave similar results. In all cases, treatment with gonadal hormones caused expected differences in body weights between control and experimental groups. No significant differences of liver weights occurred between groups within a sex.

Figure 8 shows the effects of androgen and estrogen administration to intact and castrated males on the individual sulfotransferases. Compared to controls (Figure 8A), androgen administration resulted in increased STIII levels (Figure 8B). Thus, in addition to preventing STII production, the androgen appears to elevate STIII. Castrated males given testosterone exhibited enzyme profiles similar to intact rats given the androgen. A very interesting observation was made in the study of the effect of estradiol administration to intact or castrated males (Figures 8C and D): elevated cortisol sulfotransferase activity in intact rats was due to STI and STII, while STIII was relatively unchanged. However, in castrates, all three enzymes were elevated and the enzyme profiles resembled those in females.

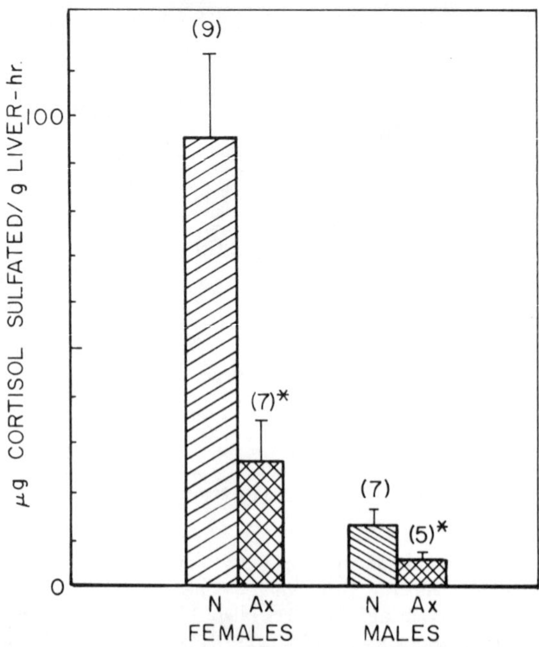

FIGURE 9. Cortisol sulfotransferase activity in liver cytosols from intact or adrenalectomized male and female rats. N and Ax represent the intact and adrenalectomized animals, respectively. The vertical lines are the standard deviations. The numbers in parentheses represent the number of individual experiments. The asterisks indicate a statistically significant difference ($P < 0.05$) between control and adrenalectomized groups within a sex.

The data support the suggested role for estrogens in the elevation of all three sulfotransferases to maximum levels and suggest that androgens control maximum STIII levels in males. In males, physiological androgen levels can apparently override the effects of pharmacological doses of estradiol on STIII. In females, testosterone administration to intact or ovariectomized animals resulted in extensive decreases of STII and the complete disappearance of STI, while STIII was not affected significantly. The data demonstrate that the androgen suppresses STI production in addition to STII production.

Although it is apparent that the gonads play an important role in the control of steroid sulfotransferase production, some nongonadal factor or factors must be invoked to explain the relatively minor effect of ovariectomy on the enzyme activity. Similarly, the failure of orchidectomy to reverse completely the inhibitory effects of androgen on STI and STII also suggests the involvement of a nongonadal factor or factors. A study to delineate the nature of these factors has been initiated.[92] The extragonadal factors appear to be of adrenal origin. As shown in Figure 9, much less cortisol sulfotransferase activity is found in the livers of 28-day-old male or female rats 30 to 50 days after adrenalectomy. Studies of the enzymatic basis for these adrenal effects are continuing, as are attempts to determine the nature of the active agents present in the adrenals.

E. Glucocorticoid Sulfates and Hepatic Enzyme Induction

Studies by Litwack and co-workers[69-71,89,90] first suggested that sulfated cortisol metabolites could be involved in hepatic enzyme induction. These researchers isolated

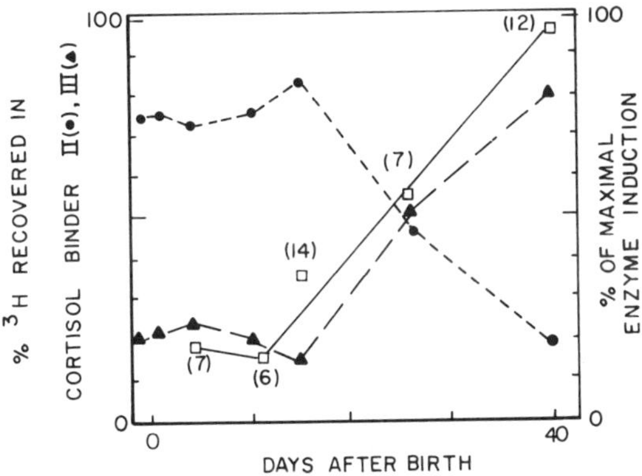

FIGURE 10. Relative amounts of radioactivity from injected ^{3}H-cortisol associated with Cortisol Binders II and III and cortisol-induced tyrosine transaminase activity in liver of rats of various ages.

four cytoplasmic proteins, Cortisol Binders, from rat liver that associated with cortisol and its metabolites after injection of physiological doses of the ^{3}H-hormone. Three of them, Binders I, III, and IV, associate with anionic metabolites[69,70] which have been identified as sulfated cortisol metabolites.[91-93] The fourth protein, Binder II,[71,123] appears to be similar to the hepatoma receptor of Tomkins and co-workers,[124-126] which associates with unmetabolized hormone and is thought to be the major determinant of glucocorticoid action.

Examination of the effects of age on tyrosine transaminase induction in relation to the Cortisol Binders by Singer and Litwack[127] suggested that the complexes between the anionic metabolites and Binders III and IV may also be involved in cortisol action. These researchers showed that the increase in cortisol-mediated tyrosine transaminase induction in growing rats parallels the increase in the amount of administered [^{3}H] cortisol found associated with Binder III. The relation between tyrosine transaminase induction and the relative amounts of Cortisol Binders II and III is illustrated in Figure 10, in which the 40-day-old animals exhibit "adult" Cortisol Binder patterns. It should be noted that although the relative percent of Cortisol Binder II complex decreases in growing animals, the absolute amount of the complex is essentially the same in rats of all ages.

The role of the sulfated cortisol metabolites is still unclear. There are differences of metabolite patterns in males and females, but males and females are equally responsive to the hormone.[127] In addition, production of maximum amounts of the single sulfated metabolite found in females, tetrahydrocortisol-21-sulfate,[91,93] preceded both "adult" Cortisol Binder patterns and maximal tyrosine transaminase induction. Inadequate production of Binders III and IV at a time when there are adequate amounts of the sulfated ligand could account for the time lag in maximal tyrosine transaminase induction. This possibility has not yet been investigated.

Additional support for the involvement of sulfated glucocorticoids in enzyme induction comes from studies of cortisol uptake, binding, and tyrosine transaminase induction in female guinea pigs carried out by Singer et al.[128] These animals exhibited much lower cortisol-mediated tyrosine transaminase induction than did female rats, despite the fact that hepatic uptake of the administered hormone would have been adequate for the maximum response in rats. Guinea pig livers contained four cortisol binding proteins that appeared similar to those in rats. However, very different Cortisol Binder patterns were found in the two species (Figure 11). Similar amounts of the complex

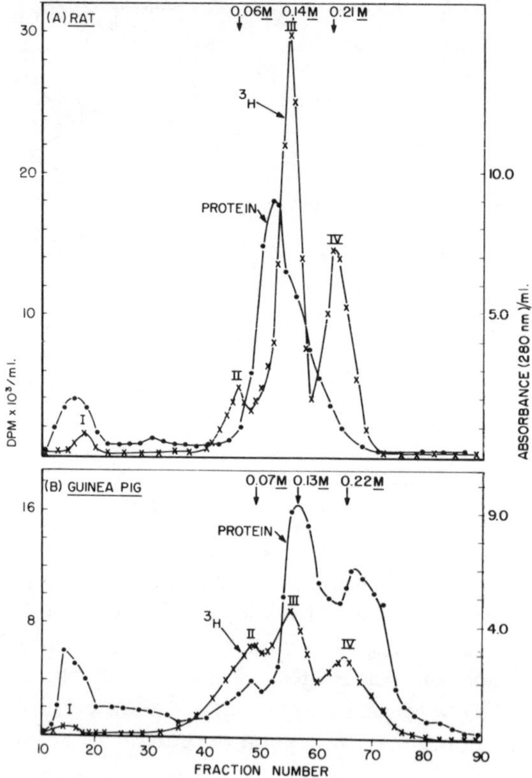

FIGURE 11. DEAE Sephadex A-50 chromatography of protein-bound radioactivity from rat and guinea pig liver cytosols 40 min after the administration of [^{3}H] cortisol. Concentrated pools of bound radioactivity, 6.0 m*ℓ* each, from (A) rat and (B) guinea pig liver cytosol were loaded on 2 × 50 cm columns of DEAE Sephadex A-50. Columns were eluted with linear gradients consisting of 300 m*ℓ* each of 0.05 *M* tris (pH 7.5) and of tris-1.0 *M* KCl (pH 7.5). Fractions of 4 m*ℓ* were collected. The positions of the peaks of steroid-protein complexes are denoted by the KCl concentrations at which they eluted from the columns. Roman numerals represent Cortisol Binders I, II, III, and IV. Recoveries were 68 to 80%. (From Singer, S. S., Gebhart, J., and Krol, J., *Eur. J. Biochem.*, 56, 595, 1975. With permission.)

between unmetabolized steroid and Binder II were observed in rats and guinea pigs. However, much smaller amounts of radioactivity from injected ^{3}H-cortisol were associated with Binders III and IV in guinea pigs than in rats. This is probably due to the fact that guinea pig liver produces only very small amounts of anionic cortisol metabolites from the injected ^{3}H-hormone and contains mostly unconjugated cortisol and tetrahydrocortisol. Singer et al.[91] have observed that there is much less cortisol sulfotransferase activity in guinea pig liver than in rat liver. These data suggest the possibility that the concentration of sulfated cortisol metabolites plus the concentration of the Cortisol Binders determine the degree of tyrosine transaminase induction.

Indirect support for a possible role for both sulfated cortisol metabolites and Binders III and IV in cortisol action comes from studies of hepatoma tissue cultures (HTC) which are used as a model of the enzyme-inducing actions of glucocorticoids.[124-126] In HTC, almost no sulfated metabolites are produced from administered cortisol. Most of the hormone is unchanged and associates with a high-affinity receptor protein that appears similar to Cortisol Binder II from rat liver.[71,123] The receptor is considered to be the determinant of glucocorticoid action, yet HTC produces only a fraction of the tyrosine transaminase induced in rat liver in response to administered cortisol.[123]

F. Glucocorticoid Sulfates in Cancer

The first evidence that glucocorticoid sulfates might be involved in cancer came from a study by LeBeau and Baulieu,[100] who demonstrated the accumulation of these compounds in adrenal tumor homogenates. More recently, several laboratories have reported that urinary glucocorticoid sulfates ("corticosteroid sulfates"), which represent the sum of cortisol, cortisone, corticosterone, and 11-dehydrocorticosterone sulfates, are elevated significantly in patients with breast, bronchial, and colonic cancers.[129-131] Ghosh et al.[129] suggested that the glucocorticoid sulfates might be of great importance in the prediction of tumor responses to adrenal ablation and other hormone manipulation. Fahl et al.[130] and Rose et al.[132] reported that the abnormal elevation of urinary glucocorticoid sulfate levels was more frequent in advanced breast cancer than in "early" tumors and that glucocorticoid sulfate levels do not necessarily correlate with the urinary free cortisol. These data suggest that the concentrations of glucocorticoid sulfates in biological fluids may prove to be a relevant index of the stage of development of certain tumors.

G. Hepatic Cortisol Sulfation and Hypertension

The adrenal cortex has long been implicated in hypertension. However, interactions between mineralocorticoids, renin, and angiotensin have most often been considered to be the sole causative agents. Recently, indications of the potential involvement of glucocorticoid sulfates in the disease have appeared in the literature. The first evidence for this possibility came from reports by Kornel and co-workers,[86,132] who observed that 17-hydroxycorticosteroid sulfates were elevated in plasma and urine from patients with essential hypertension. In a recent study, Kornel et al.[133] reported that administration of trace amounts of [^{14}C] cortisol resulted in elevated levels of cortisol-21-sulfate and cortisone-21-sulfate in plasma of patients with essential hypertension, compared to those of normotensive subjects. These observations may relate to earlier reports from several laboratories,[134-136] in which it was observed that glucocorticoids caused hypertension in rats by a mechanism that differed from the mineralocorticoid disease in that the former did not require sodium replete diets. It is possible that glucocorticoid sulfates are among the unknown steroid metabolites that several researchers suggest are involved in hypertension accompanied by depressed renin and angiotensin levels.[137,138] This possibility is supported by additional evidence. Hepp and co-workers[139] reported that the synthetic steroid 9α-fluorocortisol produced hypertension in rats in a fashion that was not dependent on dietary sodium. This hypertensive action was accompanied by decreased renin and angiotensin levels.

A preliminary report by Turcotte and Silah[101] proposed that the renal hypertension of Grollman, in male rats, was accompanied by elevation of hepatic corticosterone sulfation. A major shortcoming of this report was the fact that although statistically significant elevation of the enzyme activity was demonstrated in the hypertensive group, the authors also said that "no significant correlation was found between sulfation per mg of protein and the level of the systolic blood pressure" in individual rats. Recent studies of cortisol sulfation and Grollman hypertension, carried out by Singer et al.,[117] demonstrated that the development of increased blood pressure was parallel to the elevation of the enzyme activity (Figure 12). The insert shows that the difference in both blood pressure and cortisol sulfotransferase activity between control and experimental groups was statistically significant. No significant difference of body weight or liver weight occurred between groups.

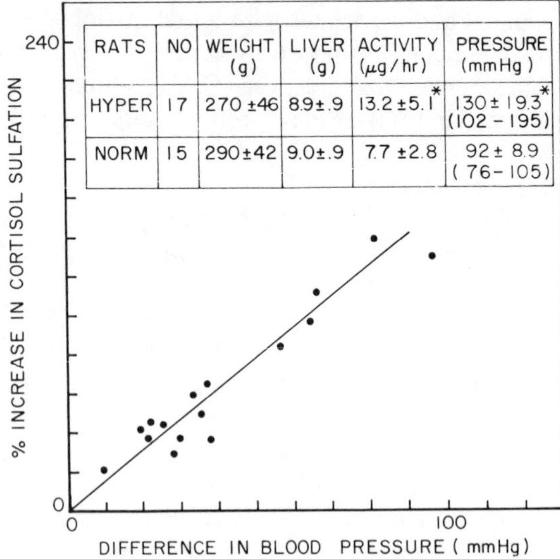

FIGURE 12. The relation between cortisol sulfotransferase activity and blood pressure in Grollman-operated male rats. Grollman-operated rats and intact controls were sacrificed and liver cytosols prepared and assayed for cortisol sulfotransferase activity. The percent increase of cortisol sulfation in hypertensives was plotted against the difference in blood pressure between control and hypertensive animals. Each point represents one experiment. The insert summarizes the blood pressure, enzyme activity, liver weight, and body weight data for the two groups of animals. The data are given as the mean ± standard deviation. Cortisol sulfotransferase activity is given as micrograms cortisol sulfated per hour per milliliter per 50% cytosol. The asterisks indicate statistically significant differences between control and experimental groups ($P < 0.01$). (From Singer, S. S., Hess, E., and Sylvester, S., *Biochem. Pharmacol.*, 26, 1033, 1977. With permission.)

Despite the correlation, the authors found Grollman hypertension a poor model for study because of the low survival rate of operated animals and the long time period — up to 6 months — needed for the development of large increases of blood pressure.[117] Therefore, several other types of hypertension were tested in hope of finding a more appropriate model.[118] Spontaneously hypertensive Okamoto rats (SHR) 16 to 17 weeks old exhibited blood pressures averaging 55 mm above those of the controls. This was accompanied by hepatic cortisol sulfotranferase levels averaging 47% above those of the controls. The hypertension elicited by daily 9α-fluorocortisol (FLHC) administration to intact rats was also examined. FLHC-injected rats developed blood pressure 40 to 76 mm higher than those of the controls after 30 to 45 days of daily injection of 0.6 mg of the hormone. This was accompanied by increases of hepatic cortisol sulfotransferase activity averaging 150%.

Fractionation of the cortisol sulfotransferase activity from Grollman, SHR, and FLHC hypertensives showed that most of the increased enzyme activity was due to STIII. A comparative chromatogram for a control animal and a FLHC-treated rat (blood pressures 90 and 165 mm, respectively) is shown in Figure 13. The data suggest the possibility that glucocorticoids, via glucocorticoid sulfates, could be hypertensive agents. In addition, they support a role for the adrenals in control of glucocorticoid sulfotransferase production.

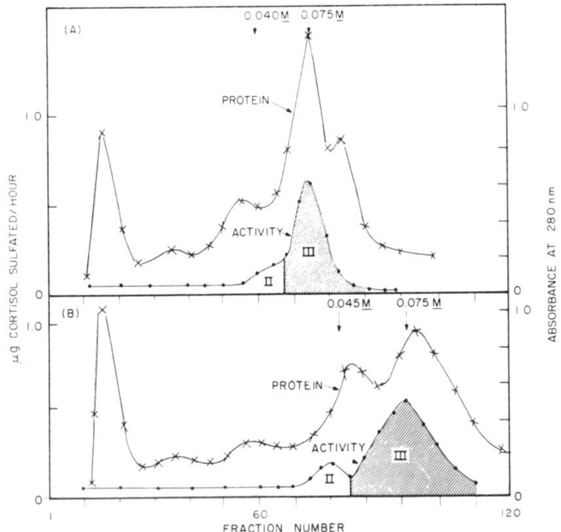

FIGURE 13. DEAE Sephadex A-50 chromatography of cytoplasmic cortisol sulfotransferase activity from livers of intact and FLHC treated male rats. The figure depicts a representative paired experiment where cytosol samples from (A) oil- and (B) FLHC-treated animals were used. Blood pressures were 90 and 165 mm, respectively. Samples (2.5 mℓ) were loaded on 2 × 50 cm columns of DEAE Sephadex A-50. The columns were then treated as described in Figure 6. STIII peaks are shaded for better comparison. Enzyme recoveries were 75%. The data are from one of four very similar experiments. (From Singer, S. S., Hess, E., and Sylvester, S., *Biochem. Pharmacol.*, 26, 1033, 1977. With permission.)

The effects of the glucocorticoids, cortisol, and corticosterone and the mineralocorticoid deoxycorticosterone on the blood pressure and hepatic cortisol sulfation in male rats were also examined.[118] These studies were carried out to determine whether the hypertensive response to these steroids, when present, could be separated from the elevation of the hepatic cortisol sulfotransferase activity. Doses of the hormones tested, which also included FLHC, that did not elevate the enzyme activity were without effect on the blood pressure. Interestingly, 3-mg doses of deoxycorticosterone caused hypertension only in rats given 1% saline as drinking water. Cortisol sulfation was elevated in this hypertensive group, but not in rats given the same dose of the mineralocorticoid and water to drink.

Fractionation of the cytoplasmic cortisol sulfotransferase activity[118] from livers of glucocorticoid hypertensives indicated that the increased enzyme activity was due mostly to STIII. Similar fractionation of the enzyme activity from deoxycorticosterone hypertensives also resulted in an enzyme profile in which the major enzyme was STIII. However, much more STII was also found in the mineralocorticoid hypertensives than in the controls.

Much more work must be done to complete our understanding of the relation between the corticosteroid sulfotransferases and hypertension. However, present data suggest that the increased enzyme activity cannot be dissociated from the development of hypertension. In addition, the studies with deoxycorticosterone suggest that there may be a relationship between glucocorticoid and mineralocorticoid hypertension that has presently been overlooked.

IV. SUMMARY

The present state of knowledge on the synthesis and metabolism of C_{21} and C_{19} steroid sulfates in mammalian tissues is reviewed in this chapter. The major emphasis is on the two classes of enzymes which determine the concentration and distribution of steroid sulfates: (1) steroid sulfotransferases that catalyze the formation of these compounds and (2) steroid sulfatases that catalyze their hydrolysis to free steroids. The chapter has been divided into two major categories of steroids, the Δ^5 — C_{21} and C_{19} 3β-hydroxysteroid sulfates and the Δ^4 — C_{21} adrenal steroid sulfates, the glucocorticoid sulfates.

Steroid sulfotransferases are soluble enzymes which are found in relatively high concentration in a variety of adult mammalian tissues. The livers of all species studied to date appear to contain steroid sulfotransferases. These enzymes are also found in adrenals and gonads of some species. The human fetal adrenal gland has a high capacity for forming neutral C_{21} and C_{19} steroid sulfates. The characteristics of steroid sulfotransferases in the different tissues are discussed.

Steroid sulfatases are membrane-bound enzymes found in many tissues. Evidence is reviewed on the possible role of steroid sulfatases in regulating the amount of free Δ^5 - 3β-hydroxysteroids available for further metabolism to hormonally active androgens or estrogens in the testis and placenta, respectively. Studies reported to date indicate the general absence of steroid sulfatases that act on glucocorticoid sulfates.

Studies in rats on pituitary control of steroid sulfatase activity of testes suggest that steroid sulfatase in the interstitial tissue may be regulated by follicle stimulating hormone (FSH), while the adrenal enzyme may be under the control of ACTH.

Investigations on cortisol sulfotransferases of the liver are discussed. The liver contains three distinct corticosteroid sulfotransferases. The concentration of these enzymes is considerably greater in female than in male rats. The production of the three sulfotransferases appears to be regulated by adrenal and gonadal steroids.

The biological significance of glucocorticoid sulfates is discussed. Reports implicating glucorticoid sulfates in hepatic enzyme induction are reviewed. Data on the relation of increased corticosteroid sulfotransferase activity and the development of hypertension are presented.

REFERENCES

1. **Vihko, R.**, Gas chromatographic — mass spectrometric studies on solvolyzable steroids in human peripheral plasma, *Acta Endocrinol. (Copenhagen) Suppl.*, 109, 52, 1966.
2. **Baulieu, E. E., Corpéchot, C., and Emiliozzi, R.**, H^3-androstene-3β,17β-diol-S^{35}-3-sulfate to H^3-dehydroepiandrosterone-S^{35}-sulfate in vivo: an example of the "direct" metabolism of a steroid conjugate, *Steroids*, 2, 430, 1963.
3. **Calvin, H. I., Vande Wiele, R. L., and Lieberman, S.**, Evidence that steroid sulfates serve as biosynthetic intermediates: in vivo conversion of pregnenolone sulfate-S^{35} to dehydroepiandrosterone sulfate-S^{35}, *Biochemistry*, 2, 648, 1963.
4. **Gasparini, F. J., Hochberg, R. B., and Lieberman, S.**, Biosynthesis of steroid sulfates by the boar testes, *Biochemistry*, 15, 3969, 1976.
5. **Payne, A. H. and Jaffe, R. B.**, Androgen formation from pregnenolone sulfate by fetal, neonatal, prepubertal and adult human testes, *J. Clin. Endocrinol. Metab.*, 40, 102, 1975.
6. **Siiteri, P. K. and MacDonald, P. C.**, The utilization of circulating dehydroepiandrosterone sulfate for estrogen synthesis during human pregnancy, *Steroids*, 2, 713, 1963.

7. *Enzyme Nomenclature,* Elsevier Scientific, Amsterdam, 1973, 189.
8. **Nose, Y. and Lipmann, F.,** Separation of steroid sulfokinases, *J. Biol. Chem.,* 233, 1348, 1958.
9. **Banerjee, R. K. and Roy, A. B.,** The sulfotransferases of guinea pig liver, *Mol. Pharmacol.,* 2, 56, 1966.
10. **Ryan, R. A. and Carroll, J.,** Studies on a 3β-hydroxysteroid sulfotransferase from rat liver, *Biochim. Biophys. Acta,* 429, 391, 1976.
11. **Adams, J. B.,** Enzymic synthesis of steroid sulphates. I. Sulphation of steroids by human adrenal extracts, *Biochim. Biophys. Acta,* 82, 572, 1964.
12. **Adams, J. B. and Edwards, A. M.,** Enzymic synthesis of steroid sulphates. VII. Association-dissociation equilibria in the steroid alcohol sulphotransferase of human adrenal gland extracts, *Biochim. Biophys. Acta,* 167, 122, 1968.
13. **Holcenberg, J. S. and Rosen, S. W.,** Enzymic sulfation of steroids by bovine tissues, *Arch. Biochem. Biophys.,* 110, 551, 1965.
14. **Fry, J. M. and Koritz, S. B.,** The effect of adrenocorticotropic hormone on sulfate metabolism in the rat adrenal gland, *Endocrinology,* 91, 852, 1972.
15. **Anderson, A. B. M., Pierrepoint, C. G., Jones, T., Griffiths, K., and Turnbull, A. C.,** Steroid biosynthesis in vitro by foetal and adult sheep adrenal tissue, *J. Reprod. Fertil.,* 22, 99, 1970.
16. **Baulieu, E., Corpechot, E. C., Dray, F., Emiliozzi, R., Lebeau, M. D., Mauvais-Jarvis, P., and Robel, P.,** An adrenal-secreted "androgen": dehydroepiandrosterone sulfate. Its metabolism and a tentative generalization on the metabolism of other steroid conjugates in man, *Rec. Prog. Horm. Res.,* 21, 411, 1965.
17. **Wieland, R. G., deCourcy, C., Levy, R. P., Zala, A. P., and Hirschmann, H.,** C_{19} O_2 steroids and some of their precursors in blood from normal human adrenals, *J. Clin. Invest.,* 44, 159, 1965.
18. **Nieschlag, E., Loriaux, D. L., Ruder, H. J., Zucker, I. R., Kirschner, M. A., and Lipsett, M. B.,** The secretion of dehydroepiandrosterone and dehydroepiandrosterone sulphate in man, *J. Endocrinol.,* 57, 123, 1973.
19. **Doouss, T. W., Skinner, S. J. M., and Couch, R. A. F.,** Synthesis of dehydroepiandrosterone sulphate by the human adrenal, *J. Endocrinol.,* 66, 1, 1975.
20. **Wallace, E. and Silberman, N.,** Biosynthesis of steroid sulfates by human ovarian tissue, *J. Biol. Chem.,* 239, 2809, 1964.
21. **Sandberg, E. C., Jenkins, R. C., and Trifon, H. M.,** Biosynthetic studies of human ovarian arrhenoblastomatous tissue in vitro. I. Sulfatase and sulfokinase activity, *Steroids,* 8, 237, 1966.
22. **Kalliala, K., Laatikainen, T., Luukkainen, T., and Vihko, R.,** Neutral steroid sulfates in human ovarian vein blood, *J. Clin. Endocrinol. Metab.,* 30, 533, 1970.
23. **Dixon, R., Vincent, V., and Kase, N.,** Biosynthesis of steroid sulfates by normal human testis, *Steroids,* 6, 757, 1965.
24. **Pierrepoint, C. G., Griffiths, K., Grant, J. K., and Stewart, J. S. S.,** Neutral steroid sulphation and estrogen biosynthesis in vitro by a feminizing leydig cell tumour of the testis, *J. Endocrinol.,* 35, 409, 1966.
25. **Pérez-Palacios, G., Lamont, K. G., Pérez, A. E., Jaffe, R. B., and Pierce, G. B.,** *De novo* formation and metabolism of steroid hormones in feminizing testes: biochemical and ultrastructural studies, *J. Clin. Endocrinol. Metab.,* 29, 786, 1969.
26. **Ruokonen, A., Laatikainen, T., Laitinen, E. A., and Vihko, R.,** Free and sulfate-conjugated neutral steroids in human testis tissue, *Biochemistry,* 11, 1411, 1972.
27. **Steinberger, E.,** Hormonal control of mammalian spermatogenesis, *Physiol. Rev.,* 51, 1, 1971.
28. **Laatikainen, T., Laitinen, E. A., and Vihko, R.,** Secretion of neutral steroid sulfates by the human testis, *J. Clin. Endocrinol. Metab.,* 29, 219, 1969.
29. **Laatikainen, T., Laitinen, E. A., and Vihko, R.,** Secretion of free and sulfate conjugated neutral steroids by the human testis. Effect of administration of human chorionic gonadotropin, *J. Clin. Endocrinol. Metab.,* 32, 59, 1971.
30. **Baulieu, E. E., Fabre-Jung, I., and Huis, in't Veld, L. G.,** Dehydroepiandrosterone sulfate: a secretory product of the boar testis, *Endocrinology,* 81, 34, 1967.
31. **Ruokonen, A. and Vihko, R.,** Steroid metabolism in human and boar testis tissue. Steroid concentrations and the position of the sulfate group in steroid sulfates, *Steroids,* 23, 1, 1974.
32. **Ruokonen, A., Vihko, R., and Niemi, M.,** Steroid metabolism in testis tissue. Concentrations of testosterone, pregnenolone and 5α-androst-16-en-3β-ol in normal and cryptorchid rat testis, and in isolated interstitial and tubular tissue, *FEBS Lett.,* 31, 321, 1973.
33. **Payne, A. H.,** unpublished data, 1977.
34. **Conrad, S. H., Pion, R. J., and Kitchin, J. D., III.** Pregnenolone sulfate in human pregnancy plasma, *J. Clin. Endocrinol. Metab.,* 27, 114, 1967.
35. **Wengle, B.,** Distribution of some steroid sulphokinases in foetal human tissues, *Acta Endocrinol. Copenhagen,* 52, 607, 1966.

36. **Huhtaniemi, I., Luukkainen, T., and Vihko, R.,** Identification and determination of neutral steroid sulphates in human foetal adrenal and liver tissue, *Acta Endocrinol. Copenhagen,* 64, 273, 1970.
37. **Jaffe, R. B., Pérez-Palacios, G., Lamont, K. G., and Givner, M. L.,** *De novo* steroid sulfate biosynthesis, *J. Clin. Endocrinol. Metab.,* 28, 1671, 1968.
38. **Huhtaniemi, I.,** Formation of neutral steroids from endogenous precursors in minced human fetal adrenals *in vitro, Steroids,* 23, 145, 1974.
39. **Cooke, B. A. and Taylor, P. D.,** Site of dehydroepiandrosterone sulphate biosynthesis in the adrenal gland of the previable foetus, *J. Endocrinol.,* 51, 547, 1971.
40. **Jaffe, R. B. and Payne, A. H.,** Gonadal steroid sulfates and sulfatase. IV. Comparative studies on steroid sulfokinase in the human fetal testis and adrenal, *J. Clin. Endocrinol. Metab.,* 33, 592, 1971.
41. **Calvin, H. I. and Lieberman, S.,** Studies on the metabolism of pregnenolone sulfate, *J. Clin. Endocrinol. Metab.,* 26, 402, 1966.
42. **Calvin, H. I. and Lieberman, S.,** Evidence that steroid sulfates serve as biosynthetic intermediates. II. *In vitro* conversion of pregnenolone-^{3}H sulfate-^{35}S, *Biochemistry,* 3, 259, 1964.
43. **Pérez-Palacios, G., Pérez, A. E., and Jaffe, R. B.,** Conversion of pregnenolone-7α-^{3}H-sulfate to other Δ^{5}-3β-hydroxysteroid sulfates by the human fetal adrenal *in vitro, J. Clin. Endocrinol. Metab.,* 28, 19, 1968.
44. **Payne, A. H. and Mason, M.,** Conversion of dehydroepiandrosterone sulfate to androst-5-enediol-3-sulfate by soluble extracts of rat testis, *Steroids,* 6, 323, 1965.
45. **Killinger, D. W. and Solomon, S.,** Synthesis of pregnenolone sulfate, dehydroepiandrosterone sulfate, 17α-hydroxypregnenolone sulfate and Δ^{5}-pregnenetriol sulfate by the normal human adrenal gland, *J. Clin. Endocrinol. Metab.,* 25, 290, 1965.
46. **Cooke, B. A., Cowan, R. A., and Taylor, P. D.,** Pathways of dehydroepiandrosterone sulfate biosynthesis in the human foetal adrenal gland, *J. Endocrinol.,* 47, 295, 1970.
47. **Siiteri, P. K. and MacDonald, P. C.,** Placental estrogen biosynthesis during human pregnancy, *J. Clin. Endocrinol. Metab.,* 26, 751, 1966.
48. **Payne, A. H., Jaffe, R. B., and Abell, M. R.,** Gonadal steroid sulfates and sulfatase. III. Correlation of human testicular sulfatase, 3β-hydroxysteroid dehydrogenase-iosmerase histologic structure and serum testosterone, *J. Clin. Endocrinol. Metab.,* 33, 582, 1971.
49. **Kawano, A., Payne, A. H., and Jaffe, R. B.,** Gonadal steroid sulfates and sulfatase. VI. Comparative metabolism in isolated seminiferous tubules and minces of human testis, *J. Clin. Endocrinol. Metab.,* 37, 441, 1973.
50. **Payne, A. H., Mason, M., and Jaffe, R. B.,** Testicular steroid sulfatase. I. Substrate specificity and inhibition, *Steroids,* 14, 685, 1969.
51. **Payne, A. H.,** Gonadal steroid sulfates and sulfatase. V. Human testicular steroid sulfatase: partial characterization and possible regulation by free steroids, *Biochim. Biophys. Acta,* 258, 473, 1972.
52. **French, A. P. and Warren, J. C.,** Properties of steroid sulphatase and arylsulphatase activities of human placenta, *Biochem. J.,* 105, 233, 1967.
53. **Hall, C. St.-G. and Giroud, C. J. P.,** Activity of human placental sulfatase and 11β-hydroxysteroid dehydrogenase with respect to steroids of the pregn-4-ene C-21-yl sulfate series, *Can. J. Biochem.,* 49, 1384, 1971.
54. **Burstein, S. and Dorfman, R. I.,** Determination of mammalian steroid sulfatase with 7α-H^{3}-3β-hydroxyandrost-5-en-17-one sulfate, *J. Biol. Chem.,* 238, 1656, 1963.
55. **Verde, A. L. and Drucker, W. D.,** Distribution of dehydroepiandrosterone sulfate sulfatase in rat tissue, *Endocrinology,* 90, 138, 1972.
56. **French, A. P. and Warren, J. C.,** Sulfatase activity in the human placenta, *Steroids,* 8, 79, 1966.
57. **Payne, A. H. and Jaffe, R. B.,** Androgen formation from pregnenolone sulfate by the human fetal ovary, *J. Clin. Endocrinol. Metab.,* 39, 300, 1974.
58. **Payne, A. H. and Kelch, R. P.,** Comparison of steroid metabolism in testicular compartments of human and rat testes, in *Hormal Regulation of Spermatogenesis,* French, F. S., Hansson, V., Ritzen, E. M., and Nayfeh, S. N., Eds., Plenum Press, New York, 1975, 97.
59. **Pulkkinen, M. O.,** Arylsulphatase and the hydrolysis of some steroid sulphates in developing organism and placenta, *Acta Physiol. Scand.,* 52, (Suppl. 180), 1961.
60. **Burstein, S. and Westort, C.,** Developmental pattern of hepatic steroid sulfatase activity in the rat, *Endocrinology,* 80, 1120, 1967.
61. **Gill, W. and Chen, C.,** Dehydroepiandrosterone sulfatase in the prostate and seminal vesicles of the rat, *Biochim. Biophys. Acta,* 218, 148, 1970.
62. **Farmsworth, W. E.,** Human prostatic dehydroepiandrosterone sulfate sulfatase, *Steroids,* 21, 647, 1973.
63. **Burstein, S.,** Studies on the solubilization and purification of rat liver microsomal steroid sulfatase, *Biochim. Biophys. Acta,* 146, 529, 1967.
64. **Notation, A. D. and Ungar, F.,** Rat testis steroid sulfatase. II. Kinetic study, *Steroids,* 14, 151, 1969.

65. **Townsley, J. D., Scheel, D. A., and Rubin, E. J.,** Inhibition of steroid 3-sulfatase by endogenous steroids. A possible mechanism controlling placental estrogen synthesis from conjugated precursors, *J. Clin. Endocrinol. Metab.,* 31, 670, 1970.
66. **Payne, A. H. and Jaffe, R. B.,** Comparative roles of dehydroepiandrosterone sulfate and androstenediol sulfate as precursors of testicular androgens, *Endocrinology,* 87, 316, 1970.
67. **Payne, A. H., Kawano, A., and Jaffe, R. B.,** Formation of dihydrotestosterone and other 5α-reduced metabolites by isolated seminiferous tubules and suspension of interstitial cells in a human testis, *J. Clin. Endocrinol. Metab.,* 37, 448, 1973.
68. **Rivarola, M. A., Podestá, E. J., Chemes, H. E., and Aguilar, D.,** *In vitro* metabolism of testosterone by whole human testis, isolated seminiferous tubules and interstitial tissue, *J. Clin. Endocrinol. Metab.,* 37, 454, 1973.
69. **Morey, K. S. and Litwack, G.,** Isolation and properties of cortisol metabolite binding proteins of rat liver, *Biochemistry,* 8, 4813, 1969.
70. **Litwack, G. and Singer, S. S.,** Subcellular actions of glucocorticoids, in *Biochemical Actions of Hormones,* Vol. 2, Litwack, G., Ed., Academic Press, New York, 1972, 113.
71. **Litwack, G., Filler, R., Lichtash, E., Rosenfeld, S. A., Wishman, C. A., and Singer, S. S.,** Liver cytosol corticosteroid binder II, a hormone receptor, *J. Biol. Chem.,* 248, 7481, 1973.
72. **France, J. T. and Liggins, G. C.,** Placental sulfatase deficiency, *J. Clin. Endocrinol. Metab.,* 29, 138, 1969.
73. **Oakey, R. E., Cawood, M. L., and McDonald, R. R.,** Biochemical and clinical observations in a pregnancy with placental sulphatase and other enzyme deficiencies, *Clin. Endocrinol.,* 3, 131, 1974.
74. **Tabei, T. and Heinrichs, W. L.,** Diagnosis of placental sulfatase deficiency, *Am. J. Obstet. Gynecol.,* 124, 409, 1976.
75. **Braunstein, G. D., Ziel, F. H., Allen, A., van de Velde, R., and Wade, M. E.,** Prenatal diagnosis of placental steroid sulfatase deficiency, *Am. J. Obstet. Gynecol.,* 126, 716, 1976.
76. **Townsley, J. D.,** Further studies on the regulation of human placental steroid 3-sulfatase activity, *Endocrinology,* 93, 172, 1973.
77. **Burril, M. W. and Greene, R. R.,** Androgenic function of the adrenals in the immature male castrate rat, *Proc. Soc. Exp. Biol. Med.,* 40, 327, 1939.
78. **Lostroh, A. J. and Li, C. H.,** Stimulation of the sex accessories of hypophysectomized male rats by non-gonadotrophic hormones of the pituitary gland, *Acta Endocrinol. (Copenhagen),* 25, 1, 1957.
79. **Notation, A. D. and Ungar, F.,** Testis steroid sulfatase activity in rats treated with chorionic gonadotropin, *Endocrinology,* 90, 1537, 1972.
80. **Georgopoulos, L. E. and Payne, A. H.,** unpublished data, 1977.
81. **Dominguez, O. V., Loza, C. A., Moran, L. Z., and Valencia, A. S.,** ACTH and sulfatase activity, *J. Steroid Biochem.,* 5, 867, 1974.
82. **Schweitzer, M., Branchaud, C., and Giroud, C. J. P.,** Maternal and umbilical cord plasma concentrations of steroids of the pregn-4-ene C-21-y sulfate series at term, *Steroids,* 14, 519, 1969.
83. **Klein, G. P., Chan, S. K., and Giroud, C. J. P.,** Urinary excretion of 17-hydroxy and 17-deoxysteroids of the pregn-4-ene series by the human newborn, *J. Clin. Endocrinol. Metab.,* 29, 1448, 1969.
84. **Pasqualini, J. R.,** Forms of urinary elimination of hormones of the cortisol group and corticosterone after administration of ACTH, *Bull. Soc. Chim. Biol.,* 45, 277, 1963.
85. **Oertel, G. W.,** On conjugation of corticosteroids in plasma, in *The Structure and Metabolism of Corticosteroids,* Pasqualini, J. R., Ed., Academic Press, New York, 1964, 65.
86. **Kornel, L. and Mitohashi, K.,** Corticosteroids in human blood. II. Free and conjugated 17-hydrocorticosteroids in essential hypertension, *J. Clin. Endocrinol. Metab.,* 25, 904, 1965.
87. **Cronhelm, T., Ericksson, H., and Gustafsson, J.-A.,** Excretion of endogenous steroids and metabolites of 4-^{14}C pregnenolone in bile of female rats, *Eur. J. Biochem.,* 19, 424, 1971.
88. **Cronhelm, T., Ericksson, H., and Gustafsson, J. -A.,** Excretion of steroid hormone metabolites in bile of male rats, *Steroids,* 19, 455, 1972.
89. **Fiala, E. S. and Litwack, G.,** Binding of a (^{14}C) hydrocortisone metabolite in liver supernatant, *Biochim. Biophys. Acta,* 124, 260, 1966.
90. **Litwack, G.,** Subcellular actions of cortisol in liver, in *Topics in Medicinal Chemistry,* Vol. 1, Rabinowitz, J. L. and Meyerson, R. M., Eds., Interscience, New York 1969, 3.
91. **Singer, S. S.,** unpublished, 1976.
92. **Singer, S. S., Morey, K. S., and Litwack, G.,** Properties of the cortisol metabolite binding system in rat liver, *Physiol. Chem. Phys.,* 2, 117, 1970.
93. **Tsong, Y. Y. and Loide, S. S.,** Isolation and characterization of cortisol metabolites from liver of adrenalectomized rat, *J. Steroid Biochem.,* 4, 239, 1973.
94. **Carlstedt-Duke, J. and Gustafsson, J. -A.,** Sexual differences in hepatic metabolism and intracellular distribution of corticosterone studied by pulse labeling with [1,2,6,7-^{3}H] corticosterone, *Biochemistry,* 14, 639, 1975.
95. **Miyazaki, M., Yoshizawa, I., and Fishman, J.,** Active O-methylation of estrogen catechol sulfates, *Biochemistry,* 8, 1669, 1969.

96. **Brooks, S. C. and Horn, L.**, Hepatic sulfation of estrogen metabolites, *Biochim. Biophys. Acta,* 231, 233, 1971.
97. **Gustafsson, J. A. and Ingelman-Sundberg, M.**, Regulation and properties of a sex-specific hydroxylase system in female rat liver microsomes active on steroid sulfates. I. General characteristics, *J. Biol. Chem.,* 249, 1940, 1974.
98. **Gustafsson, J. A. and Ingelman-Sundberg, M.**, Regulation and substrate specificity of a steroid-sulfate specific hydroxylase system in female rat liver microsomes, *J. Biol. Chem.,* 250, 3451, 1974.
99. **Ingelman-Sundberg, M.**, Specific reductive metabolism of steroid sulfates in rat liver, *Biochim. Biophys. Acta,* 431, 592, 1976.
100. **Lebeau, M. D. and Baulieu, E. E.**, *In vitro* biosynthesis of corticosteroid sulfates by adrenal tumoral tissue, *Endocrinology,* 73, 832, 1963.
101. **Turcotte, G. and Silah, J.**, Corticosterone sulfation by livers of normal and hypertensive rats, *Endocrinology,* 87, 723, 1970.
102. **Bostrom, H., Franksson, C., and Wengle, B.**, Studies on ester sulphates. XXII. Sulphate conjugation in adult human adrenal extracts, *Acta Endocrinol. (Copenhagen),* 47, 633, 1964.
103. **Klein, G. P. and Giroud, C. J. P.**, Sulfation of corticosteroids by the adrenal of human newborns, *Steroids,* 5, 765, 1964.
104. **Adams, J. B.**, Enzymic synthesis of steroid sulphates. I. Sulphation of steroids by human adrenal extracts, *Biochim. Biophys. Acta,* 82, 572, 1964.
105. **Wengle, B.**, Distribution of some steroid sulphokinases in foetal human tissues, *Acta Endocrinol. (Copenhagen),* 52, 607, 1966.
106. **Solomon, S., Bird, C. E., Ling, W., Iwayama, M., and Young, P. C. M.**, Formation and metabolism of steroids in the fetus and placenta, *Recent Prog. Horm. Res.,* 23, 297, 1972.
107. **Bostrom, H. and Wengle, B.**, Studies on ester sulphates. XXIII. Distribution of phenol and steroid sulphokinase in adult human tissues, *Acta Endocrinol. (Copenhagen),* 56, 691, 1967.
108. **Torday, J. S., Hall, G., Schweitzer, M., and Giroud, C. J. P.**, Biosynthesis of 1,2-^{3}H and 4-^{14}C steroid sulfates of high specific activity, *Can. J. Biochem.,* 48, 148, 1970.
109. **Torday, J. S., Klein, G. P., and Giroud, C. J. P.**, Influence of gonads on sulfurylation of 11-deoxycorticosterone and corticosterone by rat liver cytosol, *Can. J. Biochem.,* 49, 437, 1971.
110. **Miyabo, S. and Hisada, T.**, In vitro biosynthesis of cortisol-21-sulfate by various dog tissues, *Endocrinology,* 90, 1404, 1972.
111. **Carlstedt-Duke, J. and Gustafsson, J.-A.**, Sexual differences in hepatic sulphurylation of deoxycorticosterone in rats, *Eur. J. Biochem.,* 36, 172, 1973.
112. **Gustafsson, J.-A., Carlstedt-Duke, J., and Goldman, A. S.**, On the hepatic sulphurylating activity in male pseudohermaphroditic rats, *Proc. Soc. Exp. Biol. Med.,* 145, 908, 1974.
113. **Lipmann, F.**, Biological sulfate activation and transfer, *Science,* 128, 575, 1958.
114. **Robbins, P. W. and Lipmann, F.**, Enzymatic synthesis of adenosine-3′-phosphate-5′-phosphosulfate, *J. Biol. Chem.,* 233, 686, 1958.
115. **Wengle, B.**, Sulphate conjugation in extracts of foetal and juvenile rat liver, *Acta Soc. Med. Ups.,* 68, 154, 1963.
116. **Singer, S. S., Giera, D., Johnson, J., and Sylvester, S.**, Enzymatic sulfation of steroids. I. The enzymatic basis for the sex difference in cortisol sulfation by rat liver preparation, *Endocrinology,* 98, 963, 1976.
117. **Singer, S. S., Hess, E., and Sylvester, S.**, Hepatic cortisol sulfotransferase activity in several types of experimental hypertension in male rats, *Biochem. Pharmacol.,* 26, 1033, 1977.
118. **Singer, S. S. and Sylvester, S.**, Enzymatic sulfation of steroids. II. The control of the hepatic cortisol sulfotransferase activity and of the individual hepatic steroid sulfotransferases of rats by gonads and gonadal hormones, *Endocrinology,* 99, 1346, 1976.
119. **Singer, S. S.**, Enzymatic sulfation of steroids. III. The sulfation of corticosterone by the glucocorticoid sulfotransferases of rat liver cytosol, *Biochim. Biophys. Acta,* 539, 19, 1978.
120. **Adams, J. B. and Poulos, A.**, Enzymic synthesis of steroid sulphates. III. Isolation and properties of estrogen sulphotransferase of bovine adrenal glands, *Biochim. Biophys. Acta,* 146, 493, 1967.
121. **Wengle, B.**, Studies on ester sulphates. XVI. Use of ^{35}S-labeled inorganic sulphate for quantitative studies of sulphate conjugation in liver extracts, *Acta Chem. Scand.,* 18, 65, 1964.
122. **Roy, A. B.**, Steroid sulfatase of patella vulgata, *Biol. J.,* 62, 41, 1956.
123. **Singer, S. S., Becker, J. E., and Litwack, G.**, The principal glucocorticoid binding macromolecule in hepatoma cells in culture is similar to corticosteroid binder II of rat liver cytosol, *Biochem. Biophys. Res. Commun.,* 52, 943, 1973.
124. **Gardner, R. S. and Tomkins, G. M.**, Steroid hormone binding to a macromolecule from hepatoma tissue culture cells, *J. Biol. Chem.,* 244, 4761, 1969.
125. **Baxter, J. D. and Tomkins, G. M.**, The relationship between glucocorticoid binding and tyrosine transaminase induction in hepatoma tissue culture cells, *Proc. Natl. Acad. Sci. U.S.A.,* 65, 709, 1970.

126. **Baxter, J. D. and Tomkins, G. M.**, Specific cytoplasmic glucocorticoid hormone receptors in hepatoma tissue culture cell, *Proc. Natl. Acad. Sci. U.S.A.*, 68, 932, 1971.
127. **Singer, S. S. and Litwack, G.**, Effects of age and sex on ^{3}H-cortisol uptake, binding and metabolism in liver and on enzyme induction capacity, *Endocrinology*, 88, 1448, 1971.
128. **Singer, S. S., Gebhart, J., and Krol, J.**, Cortisol binding and tyrosine transaminase induction in guinea pig liver, *Eur. J. Biochem.*, 56, 595, 1975.
129. **Ghosh, P. C., Lockwood, E., and Pennington, G. W.**, Abnormal excretion of corticosteroid sulfates in patients with breast cancer, *Br. Med. J.*, 1, 328, 1973.
130. **Fahl, W. E., Rose, D. P., Liskowski, L., and Brown, R. R.**, Tryptophan metabolism and corticosteroids in breast cancer, *Cancer*, 34, 1691, 1974.
131. **Rose, D. P., Fahl, W. E., and Liskowski, L.**, The urinary excretion of corticosteroid sulfates by cancer patients, *Cancer*, 36, 2060, 1975.
132. **Kornel, L., Starnes, W. R., Hill, S. R., Jr., and Hill, A.**, Studies on steroid conjugates. VI. Quantitative paper chromatography of urinary corticosteroids in essential hypertension, *J. Clin. Endocrinol. Metab.*, 29, 1608, 1969.
133. **Kornel, L., Miyabo, S., Saito, Z., Cha, R. N., and Wu, F. T.**, Corticosteroids in human blood. VIII. Cortisol metabolites in plasma of normotensive subjects and patients with essential hypertension, *J. Clin. Endocrinol. Metab.*, 40, 949, 1975.
134. **Knowlton, A. I., Loeb, E. N., Stoerk, H. C., White, J. P., and Hefferna, J. F.**, Induction of arterial hypertension in normal and adrenalectomized rats given cortisone acetate, *J. Exp. Med.*, 96, 187, 1952.
135. **Friedman, S. M., Friedman, C. L., and Nakashima, M.**, The hypertensive effect of compound F acetate in the rat, *Endocrinology*, 51, 401, 1952.
136. **Knowlton, A. I., Loeb, E. N., and Stoerk, H. C.**, Effect of synthetic analogues of hydrocortisone on the blood pressure of adrenalectomized rats on sodium restriction, *Endocrinology*, 60, 768, 1957.
137. **Liddle, G. W.**, Role of the adrenal cortex in hypertension, *South. Med. J.*, 66, 51, 1973.
138. **Fukuchi, S., Nakajima, K., Takenouchi, T., and Nishisato, K.**, Plasma aldosterone in essential hypertension with low renin activity, *Jpn. Circ. J.*, 38, 1071, 1974.
139. **Hepp, R., Garbade, K., Oster, P., and Gros, F.**, Arterial hypertension produced by 9α-fluorocortisol in rats, *Acta Endocrinol. (Copenhagen)*, 75, 539, 1974.

INDEX

A

B

C

D

E

F

G

H

I

N

O

P

R

S

T

U

V

Y

Z

THE SQUIBB INSTITUTE FOR MEDICAL RESEARCH
LIBRARY
OCT 1 5 1980
PRINCETON, N. J.